Environmental Intimacies from India's North East

This book explores the intimate relationship of the people of North East India with the environment as evinced from their traditional beliefs, cultural practices, and livelihood. It offers insights into the challenges and adaptability of communities through actions ensuring conservation of the environment yet sustaining their livelihood.

The book encompasses movement toward responsible actions and the engagement of communities in pro-environmental behavior. It showcases the various practices, knowledge, and debates on the environment-behavior relationship of the people of North East India. It brings into focus some of the daunting environmental issues and the endeavors for sustenance that is true to a region which is rich in ecological diversity and touches the spirit and psyche of its people.

This volume will be of interest to scholars and researchers of psychology, anthropology, sociology, environmental sciences, development studies, law and governance, environmental administration, and South Asia studies.

Indranee Phookan Borooah teaches in the Department of Psychology at Gauhati University, Guwahati, Assam, India, as a professor. During her long career her focus in research has been positive psychology. The positive social and cultural experiences of people have attracted curiosity in her, especially in the context of positive cultural psychology of North East India. Her current research is on some cultural aspects of Assam. She has been awarded for her contributions to school psychology in India, with special focus on North East India, and for teaching, research, and administration.

Bidita Das is an assistant professor in the Department of Psychology at Handique Girls' College, Guwahati, Assam, India. She is a certified Positive Psychology Practitioner from School of Positive Transformation, USA. Her core research interest lies in positive psychological aspects of human behaviour. She is involved in several non-governmental organizations working in the areas of mental health and youth

Sabiha Alam Choudhury teaches in the Department of Psychology Mizoram University, Aizawl India as an assistant professor. Her area of specialization is positive psychology and marriage and family counselling. She is a certified CBT therapist and motivational interviewing trainer. She has been conducting training and awareness programs in several areas in North East India and was awarded the best performance award for extraordinary contribution to school psychology in India.

Environmental Intimacies from India's North East

Psycho-Social Implications for Pro-Environment Behaviour

Edited by Indranee Phookan Borooah, Bidita Das and Sabiha Alam Choudhury

Routledge
Taylor & Francis Group
LONDON AND NEW YORK

First published 2025
by Routledge
4 Park Square, Milton Park, Abingdon, Oxon OX14 4RN

and by Routledge
605 Third Avenue, New York, NY 10158

Routledge is an imprint of the Taylor & Francis Group, an informa business

British Library Cataloguing-in-Publication Data
A catalogue record for this book is available from the British Library

ISBN: 978-1-032-82506-9 (hbk)
ISBN: 978-1-032-82507-6 (pbk)
ISBN: 978-1-003-50487-0 (ebk)

DOI: 10.4324/9781003504870

Typeset in Sabon
by SPi Technologies India Pvt Ltd (Straive)

Dedicated to our Fathers

Prafulla Nath Phookan & Nagen Das

(In Memory)

from whom we imbibed a love of nature

And

Alamgir Chaudhury

whose affinity to nature is inspirational

Contents

Figures

Tables

Contributors

Amri, Tiasunep, Ph.D. is an assistant professor in the Department of Anthropology, Nagaland University, Zunheboto, Lumami, India. Currently his research focuses on the interface of cultural values and economics. Some of his other interests are evolutionary biology and etymology of Tibeto-Burman words.

Arunkumar, M. C., is a professor in social and cultural snthropology in the Department of Anthropology, Manipur University, India. His areas of interest are social problems, social psychology and gender studies. He has a number of papers and books to his credit. He is associated with All India Radio, Imphal, and local cable networks as a news analyst.

Baglari, Happy, is an RCI Registered Clinical Psychologist and assistant professor of Centre for Studies in Applied Psychology at Dibrugarh University, Assam. She is currently pursuing her Ph.D. in psychology from Gauhati University, Assam. She has experience both at industry and academic level and is professionally trained in psychometric assessment and personality mapping, and the psychotherapies cognitive behavior therapy and stress management.

Barman, Mayuri, Ph.D. is currently an assistant professor (Grade III) in Pandu College, Department of Philosophy, Guwahati Assam. She has several research publications (both national and international) to her credit. She was a Gold Medalist in Philosophy in the year 1997–1998 Gauhati University (She received a Research Excellence Award. She was also the recipient of the Achievers Icon Award in "Outstanding Excellence in Philosophy" in the year 2020 by Brand Opus.

Begum, Jesmin, At present she is working as a faculty in Social Work Department in Mason Phillips Academy, Tura and pursuing her Ph.D from Martin Luther Christian University, Shillong. She has recently completed a project on implementation research to assess the capacity of the Health and Wellness Centres to deliver comprehensive primary health care services in the state of Meghalaya.

Bhattacharjee, Biswendu, has been in active journalism for the last 25 years in Tripura and worked with many national, regional, and local newspapers. At present he is serving for *Times of India* and the largest wire news service, United News of India. He is a well-known environmental personality of North East for his carbon neutral lifestyle and eco-activism. He is the Chairman of Public Relation Council of India, Tripura Chapter.

Bhattacharjee, Somenath, is currently a professor in the Department of Anthropology, Nagaland University. His area of specialization is Social-Cultural Anthropology. His areas of research are Tribal Ethnography, Medical Anthropology, Folk and Material Culture. He has published about 80 research papers and articles in National,International Journals and Edited volumes. He has authored three books and he has been associated with four research projects, funded by UGC, ICMR, ICSSR-NERC, ICSSR-IMPRESS.

Borah, Annesha, Ph.D., is currently working as an assistant professor in the Department of Geography at Assam Royal Global University in Guwahati, Assam, India. She has published in several international and national journals. Currently she is working on a TRIBAC–funded research project on 'Current Status and Development Potential of Bamboo Industry in Assam.'

C. Zothanmawia is an assistant professor at the Department of Psychology, Mizoram University. His expertise lies in areas such as acculturation, psychological analysis of disasters, and the effects of socio-economic factors on acculturative stress.

Chaudhury, Runti, Ph.D., is an assistant professor in the Department of Geological Sciences, Gauhati University. She was the recipient of the Fulbright-Nehru Doctoral Research Fellowship (2014–15) and conducted a part of her research work at Lamont Doherty Earth Observatory of Columbia University, New York.

Deka, Jyoti Prakash, Ph.D., is teaching, researching, and supervising master's students in the Department of Environmental Science, Gauhati University, as a teaching associate. His interest in teaching is highly evident from his keen interest in many environmental issues other than his own area. Jyoti Prakash Deka has published four international journal papers on his Ph.D. work and co-authored ten other international journal papers

Droma, Pema, is a guest assistant professor of English at Central Institute of Himalayan Culture Studies, which is an autonomous body under the Ministry of Culture of India. She has worked on several projects under the NGO Flowering Dharma and is one of the co-founders of the NGO Arunachal Mindfulness.

Duari, Mousum is currently working as a project associate at Advanced Diploma in Child Guidance and Counselling in National Institute of

Public Cooperation and Child Development (NIPCCD), Guwahati and as a mental health counsellor at Adhyaya (Luit Fellows), Guwahati. She has worked as a PGT psychology teacher cum counsellor at NPS International School, Guwahati and prior to that as a psychology lecturer at undergraduate level at Assam Down Town University, Guwahati.

Ete, Leeyir, is a guest assistant professor in the Department of Sports Psychology, Rajiv Gandhi University, Arunachal Pradesh, India. She is the present chairperson of the All Arunachal Pradesh Psychological Association and has organized outreach programs and provided telecounselling services.

Jamir, Latonglila, Ph.D., teaches in the Department of Environmental Science at Nagaland University. Her research interests lie in areas such as development of advanced materials for water remediation; environmental monitoring and management; green synthesis of organic compounds and biological assessment of the compounds thus synthesized; and more.

Ketiyala completed her master's in environmental science from the Department of Environmental Science at Nagaland University. She is currently a research scholar in the same department.

Kikon, Lobeni, is currently working as a faculty in Social Work Department, Mason Phillips Academy, Tura, Meghalaya and also pursuing her Ph.D. She has worked as a counselor at Civil Hospital, Dimapur, Nagaland, India. She has completed a project on Implementation Research to Assess the Capacity of the Health Wellness Centers.

Lahkar, Baharul Islam, is Chair-Professor, PA Sangma Centre for Tribal Studies, University of Science & Technology Meghalaya (USTM). He has more than 25 years of experience in research, teaching, and training. He is the chief editor of *Vistas de Society* (Journal of the School of Social Sciences, USTM) and member of professional bodies like Indian Sociological Society, Indian Social Science Academy, NE India Education Society, Assam State Child Protection Society, and Assam Sahitya Sabha.

Laldinpuia is currently a research scholar at the Department of Psychology, Mizoram University (MZU). With previous experience as a guest faculty at the Department of Psychology, Pachhunga University College Campus, Mizoram University. He harbours a special interest in studying the intercultural intricacies of the Kuki-Chin-Mizo people, recognizing the importance of exploring one's own cultural background in psychological research.

Lotha, Tsenbeni N, acquired a master's degree in chemistry and is presently pursuing her Ph.D. degree at the Department of Environmental Science at Nagaland University. Her research area is on adsorption-based wastewater remediation studies.

Nakro, Vevosa, is a research scholar in the Department of Environmental Science at Nagaland University. He completed his master's in environmental science in the same department and his research interest is in development of advanced materials for wastewater remediation.

Ralte, Rinpari is currently serving as Professor and Head of the Department of Psychology at Pachhunga University College Campus, Mizoram University. She has translated a book, *School Psychology in the Indian Context*, authored by Prof. Panch Ramalingam in the Mizo language and has been awarded Best School Psychology Trainer Award and InSPA Rathna Award by Indian School Psychology Association for her continued efforts in the promotion of school psychology in her community.

Rongpi, Bamonkiri, Ph.D., has worked in the area of health, disease, and various forms of treatment which are practiced and followed by the Karbi people of Karbi Anglong district of Assam. With a research experience of more than five years, he is also a member of the Centre for Karbi Studies (CKS) Diphu, Assam, India. Currently he is working as a teacher of anthropology in Renu Terangpi Memorial College Diphu.

Shangpliang, Rekha M. is a professor in the Department of Sociology, North East Hills University. She has three books to her credit and has authored a number of articles on matriliny, gender, and environment related themes in national and international journals.

Sharma, Pallavi, Ph.D., is an assistant professor in the Department of Environmental Sciences, Gauhati University. Pallavi Sharma has been working with drinking water and ground water quality in rural areas of Assam. She was working with UNICEF in a joint venture for capacity building for determination of household water quality and its handling with relation to the Water and Sanitation Programme (WASH) of the government of India.

Foreword

Dr Anwaruddin Choudhury, PhD, DSc, IAS (retd)
Member, IUCN/SSC/BLI Specialist Groups, and
State Board for Wildlife; retd. Divisional Commissioner
and Commissioner & Secretary, Assam

Environmental issues have taken centre stage in recent years owing to various reasons. Continuous onslaught of humans over nature has resulted in extinction of various species and their habitat. With a growing human population, the biotic pressure is on the rise despite a growing awareness. Traditional values, which used to give forest and wildlife much importance, are on the wane. After all, one must realise conservation is not just saving some forest or some species but is for the survival of humankind. Species are indictors of ecological balance, and the forest cover is essential for meeting the needs of the people as well as the climate, including fresh water.

In this book, entitled *Environmental Intimacies from India's North East: Psycho-Social Implications for Pro-Environment Behaviour*, the editors have carefully selected some essays that inform people that there are some good examples still being followed in the lesser-known areas of India's North East. Straight from folk culture to traditional belief and rituals, all have some space for environment and community conservation. These are necessary to remind those who have digressed and to others to replicate such good examples. The book is edited by Indranee Phookan Borooah, Bidita Das, and Sabiha Alam Choudhury, who have made their best effort to portray most of the states of North East India.

(A.U. Choudhury)

Anwaruddin Choudhury is an Indian naturalist, noted for his expertise on the fauna of North East India and its adjacent areas. Choudhury is an ornithologist, mammalogist, artist, civil servant, photographer, and author. He is a trustee and the Honorary Chief Executive of the Rhino Foundation for nature in North East India, was Deputy Commissioner (District Magistrate) of Baksa and Lakhimpur districts in Assam and also served as Secretary with the government of Assam. He retired as Divisional Commissioner of Barak valley, Development Commissioner for Hill Areas, and Commissioner & Secretary to the government of Assam. Known as the "Birdman of Assam," he was the first to produce books on the birds of different North Eastern states including Assam, Arunachal Pradesh, Nagaland, Mizoram, and Meghalaya. His studies have contributed enormously to conservation and awareness in North East India. He is the author of 28 books and more than 950 articles and scientific papers.

Preface

Growing up amidst the verdant environment of Assam (a state known for its flora and fauna and the mighty Brahmaputra as the lifeline) and travelling as professionals and leisure travellers to the various states of the North East region of India, we, the three editors, have been struck by the pristine beauty of this region with its many-shaded green landscape of hills and valleys and a sky frequently changing colour from a bright azure to a furious dark grey, followed by heavy downpours of sheets of rain, and then the emerging of a freshly washed vista of greenery shining with raindrops. Moreover, the people who are so diverse yet connected by their close relationship with this beautiful natural environment, with which they share a special intimacy because of its role in every aspect of their life (be it livelihood, spiritual practices, social rituals, or their creative pursuits such as their weaving, singing, dancing, painting, or craftsmanship), have struck us with awe and made us ponder on the possibility that closeness with nature brings out the best in human endeavour and creativity.

The North East of India is diverse and its diversity adds on to its uniqueness. As proud residents of the North East of India, we take immense pride in this region as being the greenest, with the last untouched places, in the country. The lush natural environment has a lot to do with the indigenous practices that are still performed in the form of rituals or festivities round the calendar. These have a close connection to either the harvest or praying to the nature Gods for its bounty. As researchers and teachers, we have many conversations regarding the world and the people around us. On one such conversation, we happened to ponder deeper and deeper into the idea of bringing into light the pro-environmental habits of the people seen in various corners of our region. This conversation has taken the shape of this compiled book in which we have tried to put forward the present scenario, the intimate aspects of the man-environment relationship seen in various forms in the states of the North East and then highlighted how these intimacies have been worked upon and applied in many forms, across the North East for conserving the environment.

Every morning as we take newspapers in our hands or scroll through the news apps, we find at least one story that covers the state of the environment around us. Be it the unexpected rise in temperature in places that were among the cooler spots of the country, the delayed or short winter, incessant rain, prolonged dry spell, or the parched condition of the earth. Normal conversations among friends and family frequently include the topic of climate change! Be it on global platforms or tiny households, everyone acknowledges the fact that climate is changing for the worse. As teachers in different educational institutes of the region, we have witnessed students' activism in promoting and creating awareness about the global climate scenario. Our institutes have been installed with climate clocks that remind us how the ticking climate clock is our collective responsibility. There are alarms set off by scientists, naturalists, world organizations such as the United Nations, and others. As cited in the Hindu on 11 April 2024, Simon Stiell, Executive Secretary of the United Nations Framework Convention on Climate Change, has said that the next two years are "essential in saving our planet." He further said that the leading economic powers need to step up to decrease greenhouse gas as current efforts are hardly effective to stabilise temperature rise and climate change.

There is enough evidence of rising temperatures, extreme weather events, and their consequences on ecosystems and communities across the globe. The situation in the North East region of India is no different; however, there is hope that if steps are taken immediately to halt the mindless misuse and exploitation of the natural environment and actions are taken towards conservation, the pristine environment of this region can be restored. Olson and Dinerstein (2002) have stated that Arunachal Pradesh "is one of the 200 globally important eco-regions of the world where conservation is expected to achieve the goal of saving a broad diversity of Earth's ecosystems and ecological processes."

"Whilst various studies report that knowledge regarding the destruction of the planet and the sensitivity toward the improvement and protection of the environment have increased among the general population, this does not appear to have translated into specific behaviours on the part of society" (Álvarez and Vega, 2009). However, as evidenced by psychologists it is the emotional connection and the intimate bonding with one's environment which directs pro-environment behaviour. The communities in the North East already exhibit such an intimacy as seen in their daily lives in the form of household use of natural resources, their livelihood, their spiritual rituals and practices, their myths and folklore, and so on, which are all reflected in the chapters of this book.

The goal to achieve climate stability will not be possible until communities rise together for pro-environment behaviour modification. And behaviour modification is impossible if people do not have an emotional connection attached to the change that is desired. Something needs to stir up within the consciousness of the collective society as mentioned in the chapter on Oikophilia. That nature gives back to survival and sustenance is seen in the chapter on Manas Tiger Reserve. The chapter on Selbagre Wildlife Sanctuary reiterates the fact that community initiatives can hugely influence the community to re-work on conservation practices. The degradation of the environment is a result of the adventures of 'man,' and it is the same 'man' who needs to wake up and take conscious pro-environment decisions for its existence.

This book is our honest attempt in compiling the various practices that showcase the man-environment relationship which is being nurtured and maintained in the North East of India. We have brought together the various cultural and psychological strings that connect us with nature, and which have contributed to its conservation. The authors of the chapters of this book have presented various facets of pro-environmental practices and how these could be utilised to imbibe in the youth and future generations environmental intimacies, pro-environment behaviour, and conservation of the environment.

For assisting us in putting this book together we would like to acknowledge Dr. Anwaruddin Choudhury, Amlan Baruah, and Jayanta Borooah for the support received from them. We particularly acknowledge every author who has represented their states and put forward aspects of their culture and people which reflect their intimate relationship with the natural world.

Indranee Phookan Borooah
Bidita Das
Sabiha Alam Choudhury
30th April, 2024

Part I

Ecological Status Today

1 Way We Are

Introduction

Bidita Das, Sabiha Alam Choudhury and Indranee Phookan Borooah

The news of the four Columbian children who survived in the dense forests of the Amazon for 40 days caught everyone by surprise in the month of June 2023. It is no less than a miracle. But a deeper understanding of the incident would unravel the skills of survival possessed by these children. These skills are closely attached to the knowledge of the natural world around them. It is reported that the Columbian children were trained from a very early age to understand the nature around them; this meant they were always close to the jungles, knew what fruits were edible and what to avoid. It is also reported that the eldest child among the four, aged 13, fed the youngest 11-month-old using tree leaves as a drip.

There have been worldwide discussions on multiple forums regarding the emergency that climate change is causing. Instances of floods, avalanches, extreme dry spells and forest fires have filled up the internet. Every country on this planet is facing the brunt of the massive climate turn. But is humanity prepared? It is a pertinent question that needs to be addressed.

Human activities are the primary cause of this massive shift in climatic conditions. It is now well established that the magnitude and rate of contemporary warming of the atmosphere, land and oceans are predominantly due to human activities such as the burning of fossil fuels, deforestation and changes in land use and land cover (LULC) during the industrial period, which has substantially changed the atmospheric composition of greenhouse gases. As per the reports of the United Nations, 2010–2019 was the warmest decade recorded, which brought wildfires, hurricanes, droughts, floods and numerous climate disasters across the globe. It has been reported that between 2010 and 2020, highly vulnerable regions which are home to approximately 3.3–3.6 billion people experienced 15 times higher human mortality rates from floods, droughts and storms compared to regions with very low vulnerability.

The impact of climate change has social implications. Climate change can aggravate storms and disasters leading to threats such as food and water scarcity, which in turn can lead to conflict, scarce resources, market fluctuations, forced human migration, infectious diseases and so on. There has been much happening around the world; thrust on investments for renewable energy have increased manifold. But is this enough? The world took a significant first step

DOI: 10.4324/9781003504870-2

by adopting the Paris Agreement in December 2015, which was adopted by 196 parties at the UN Climate Change Conference (COP21) in Paris, in which all countries had committed to take action to address climate change. More actions, however, are critically important to meet the targets.

Intergovernmental negotiations to address climate change have been present for three decades now. India has been an active player in this process right from its inception. At the Conference of Select Developing Countries on Global Environmental Issues convened in New Delhi in April 1990—the first of its kind for developing countries—India succeeded in securing the general support of the developing world for its basic international positions on climate change. These were first, that the primary responsibility for reducing greenhouse gas (GHG) emissions causing the problem of climate change rested with the developed world since they were the ones responsible for producing the bulk of these emissions. Second, the emissions of developing countries were still very low and as they needed to grow to meet their future development and poverty reduction needs, hence no GHG reduction targets could be prescribed for them. Third, any formal agreement on climate change needed to provide for technology transfer and funds for developing countries to help them address this challenge (Ministry of Environment and Forests, 1990).

India's role in shaping the background against which the convention negotiations were held cannot be overlooked. The original draft of the First Assessment Report of the Intergovernmental Panel on Climate Change (IPCC) had noted that both developed and developing countries had "common responsibilities" on climate change. India worked closely with other developing nations to ensure that this was amended to "common but differentiated responsibilities" (CBDR) of industrialized and developing countries (Sengupta, 2019a). India played a key role in ensuring that the convention negotiations were undertaken through an Intergovernmental Negotiating Committee operating under the direct authority of the United Nations General Assembly (UNGA)—to allow for "openness, transparency, universality and legitimacy" and the "full participation" of all States (World Meteorological Organization [WMO]/United Nations Environment Programme [UNEP], 1990)—rather than through other specialized forums, such as UNEP or IPCC, which were being advocated by developed countries at that time.

Climate Change

A significant issue for the environment and human progress is climate change. Climate change is the considerable alteration of average weather conditions over several decades or more, such as circumstances becoming warmer, wetter or drier. The difference between climate change and natural weather variability is in the longer-term shifts. The climate of the Earth varies from location to location based on latitude, vegetation, mountains and seasonal variations. Therefore, climate change refers to a variation in the average climate that lasts for a long time (decades or longer) in each location or region.

The world's industrial revolution at the turn of the 19th century started to alter the atmosphere of the planet. According to the United Nations Framework Convention on Climate Change (UNFCCC), climate change is the change in climate that is attributable directly or indirectly to human activity that modifies the composition of global atmospheric gases since the beginning of the industrial revolution.

Climate system (temperature, precipitation, sea level, ice cover, etc.), natural resources (ecosystems, biodiversity, land, water, etc.), human security (livelihood, food, energy and health), and social stability (migration, violence, conflict, etc.) are strongly interconnected and provide feedback to one another (Scheffran et al., 2023). One of the key determinants for the world's many ecosystem activities, processes and services is the climate. The effects of climate change will have a grave influence on people's daily lives, including their access to water, food, shelter and other necessities (IPCC, 2014). In areas/regions where people are more reliant on natural resources, the effects will be more severe. The effects of climate change on natural resources are widespread.

India and Climate Change

India recognizes that climate change is a significant global problem and an issue of great socio-economic and political importance. India's participation in the global climate discourse has been shaped primarily by a scientific understanding of its domestic vulnerabilities to climate change and the adverse impacts on the country's environment, economy and society. India has a distinct geographical entity (Indo-Gangetic Plains, Central Highlands and Deccan Plateau, Thar Desert, Himalaya and North East Mountain Ranges, East Coast, West Coast, and bordering seas and islands) with significant variations in climate and regional and local weather conditions across the country. Given its varied geographical terrain, it is one of the most disaster-prone countries in the world, "*where 59% of the land is vulnerable to earthquakes, 8.5% of the land is vulnerable to cyclonic storms and 5% of the land is susceptible to river basin floods*" (Maritime India.org). Moreover, much of the population is dependent on climate-sensitive sectors like agriculture, forestry and coastal ecosystems for livelihoods. Climate-change-induced extreme temperatures, heavy precipitation, cyclonic storms and sea level rise exacerbate India's vulnerability to a whole range of hydrometeorological hazards which are rising alarmingly, both in terms of their intensity and frequency, and decreasing their predictability.

According to a 2020 report by the Ministry of Earth Sciences (MoES) of the government of India titled "Assessment of Climate Change over the Indian Region," summer heat waves are projected to increase by three to four times in frequency and are expected to become longer in duration by the end of the century compared to the 1976–2005 baseline period. The report noted that India's summer monsoon precipitation from June to September has reduced by 6% as compared to the period 1951 to 2015, with a marked decline in rainfall over the Indo-Gangetic Plains and the Western Ghats. At the same time, more intense extreme wet spells during the summer monsoon season have been observed over central India. Specifically, the intensity of rainfall beyond 150 mm per day increased by about 75%during 1950–2015 and the frequency of daily precipitation extremes also increased during this period.

India no doubt has taken a leadership role in driving global action on climate change. It used indigenous technology to enhance its resources and promote green energy to reduce carbon emissions. Alongside this, India has co-founded the International Solar Alliance (ISA) with France, and in doing so, is leading the global movement towards solar power, with a focus on promoting energy access and transition. Already, the ISA has 110 member countries and is pursuing nine programmes promoting 10 GW of off-grid and grid-connected solar projects in developing countries. The one sun, one world, one grid project (OSOWOG) along with the United Kingdom is based on the vision of building and scaling inter-regional energy grids to share solar energy across the globe. It can be the solution to most of the global problems in the energy sector. India's pledge to reach net-zero emissions by 2070 was one of the most important announcements at COP26.

India's source of energy mix is carbon intensive. Coal is its primary source of fuel, accounting for approximately 70% of electricity generation. It powers much of its transport through oil. As a result, India is the world's third-largest emitter of CO2, despite low per capita CO2 emissions. However, India has a unique opportunity to turn to no-emissions technology as India is the world's third-largest producer of solar energy and enjoys the lowest renewables costs; it has also pioneered in green hydrogen (hydrogen made using renewable energy).

In 2008, India introduced the National Action Plan on Climate Change (NAPCC). The NAPCC addresses the urgent and critical concerns of the country through a directional shift in the development pathway. The NAPCC identifies measures for addressing climate change effectively. It outlined several steps to advance India's development and climate change–related objectives of adaptation. Eight National Missions, which form the core of the National Action Plan, represent integrated, multi-pronged and long-term strategies for achieving key goals in the context of climate change. These are: National Solar Mission, National Mission for Enhanced Energy Efficiency, National Mission on Sustainable Habitat, National Water Mission, National Mission for Sustaining the Himalayan Ecosystem, National Mission for a Green India, National Mission for Sustainable Agriculture and National Mission on Strategic Knowledge for Climate Change.

The NAPCC recommended that each state should develop its own Action Plan for Climate Change (SAPCC), which addresses the state-specific vulnerabilities due to climate change and captures these through state-specific adaptation strategies while contributing to the national mitigation efforts. This approach was adopted keeping in mind the geographic diversity in the country and initiatives that have been taken up by the states and its communities. Thirty-four states/union territories (UTs) prepared their SAPCC in line with NAPCC considering the state-specific issues related to climate change. Even though this was launched in 2008, reports have failed to document the success of the NAPCC.

India's North East

A separate climate change mitigation programme for India's North East, the North East Climate Change Adaptation Programme (NECCAP) was aimed to support the government of India. The purpose of the programme was to strengthen the adaptive capacities of target groups, reducing their vulnerability to climate change in a target-oriented manner and facilitate adaptation to climate change in the North East Region (NER) of India. The design of the programme is based on the overall policy of the government of India on climate change and on agreements made between Ministry for DoNER and the state governments of Assam, Nagaland, Meghalaya and Sikkim. The programme was envisaged to contribute to the Indian NAPCC as well as to the related missions (i.e. Green India Mission, Mission on Himalayan Ecosystems, National Mission on Strategic Knowledge on Climate Change, Mission on Sustainable Agriculture, and National Water Mission). The programme was aligned to the priority areas of the respective SAPCC, and to the National Policy on Climate Change, and was closely coordinated with the Ministry of Environment and Forests.

The project activities ranged from policy advice at the state level to direct community level support and the creation of partnerships and networks in the region and beyond. Additionally, the project applied a range of human, institutional and societal capacity development measures like:

- Preparation and assistance in implementing State Action Plans on Climate Change
- Drafting of a water policy in Meghalaya
- Spring-shed development and the preparation of village water security plans in Tendong Hill region, Sikkim
- Support for the cultivation of indigenous rice varieties in Meghalaya and Nagaland
- Improvement of Eri silk production in Assam and Meghalaya
- A study on the regeneration, conservation, and sustainable management of oak forests in Sikkim
- Inputs for the integrated management of land and water resources in the Upper Umiam River Basin, Meghalaya
- Breeding of indigenous fish species in Meghalaya
- Support for the compilation of climate change–relevant data in Meghalaya
- Training and awareness raising on climate change, at government and community levels

The evaluation of these government initiatives over the years has not yielded sustainable results. The implementation of these projects did not deliver what it proposed to deliver. As reported in the *Indian Journal of Occupational and Environmental Medicine*, these initiatives to prevent climate change did start but were not continuous and sustainable. By releasing the NAPCC, the government did show India's commitment to address climate change issues and sent a positive message to the public, industries and civil society about the government's concern to address the climate change issue through concerted action. Issues related to the awareness regarding global warming and climate change among the general population and the issue related to agriculture and health hazards due to climate change still need to be addressed more seriously (Pandve, 2009).

Effects of Climate Change on Indigenous Population

Global climate change has a significant impact on natural resources. Natural disasters, unusual weather patterns, groundwater depletion, poor crop yields, fish population declines and an increase in tropical diseases are currently the main risks to people (Adger et al., 2014). If people are entirely dependent on natural resources, for instance, the catastrophic weather occurrences brought on by climate change would likely force them to migrate (IPCC, 2014).

Indigenous people are uniquely impacted by both the effects of climate change and the policies and measures taken to combat it. Indigenous people are crucial to the effectiveness of policies and actions aimed at mitigating and adapting to climate change, as well as to just transition policies as workers. Their situation is distinct from that of other communities and from that of the impoverished, given the size and complexity of the risks they confront related to climate change, including specific threats to their livelihoods, customs and ways of life. Indigenous people, with their traditional knowledge and livelihoods, have a special role to play in climate action, spanning both efforts at adaptation and just transition policies.

Deeply ingrained spiritual, cultural and economic linkages between indigenous tribes and the environment exist. However, the livelihoods of indigenous groups are being disrupted in various tropical nations by rising temperatures, droughts, floods, heat waves, wildfires, landslides and other extreme weather events, which limits their capacity to deal with and adapt to climate variability. Their food sovereignty is declining because of crop failure and increased hunger. Therefore, indigenous peoples must be seen as powerful agents of change, given access to decent employment opportunities, and allowed to take part in the creation, implementation and evaluation of sustainable policies and measures aimed at combating climate change if climate action is to be successful. The reasons that render indigenous peoples particularly vulnerable to climate change must be addressed in a specific and focused way.

Tribal populations occupy over 20% of the Earth. They are the planet's most vital guardians because of their extraordinary kinship with the soil, water and other natural resources. Seven states make up North East India, which is located at the meeting point of South East Asia and India. The region's distinctive features, which set it apart from other areas of the Indian subcontinent, include the verdant environment, geological and ecological diversity, and the exotic range of communities. The region's indigenous group, which makes up 27% of the total population and has distinctive ethnic and cultural traits, plays a major role in forming the region's physical characteristics in a geological sense. Each state boasts scenic hills and verdant meadows that are home to thousands of different types of flora and fauna, making it a traveller's paradise.

In the eyes of the indigenous population of the North Eastern Region of India, happiness and biological resources go hand in hand. They demonstrate their symbiotic affinity with nature through their ancient forms of worship. There is a high degree of coexistence between numerous customs, festivals and religious rituals, traditional values and beliefs, livelihood practices and trade, traditional administrative and financial management strategies, and so on, and the serene, resilient and joyous energy of nature.

The People and Community Collaboration

The North Eastern Region of India, which is tucked away in the Himalayas, consists of Sikkim as well as the "Seven Sister States" of Arunachal Pradesh, Assam, Manipur, Meghalaya, Mizoram, Nagaland and Tripura. In addition to being home to numerous distinct ethnic groups with a rich cultural legacy and traditional understanding of the environment, the region is a hotspot for biodiversity on a global scale.

India is a large country with interspersed legacy and identity largely driven by its people and their natural habitats. The country has 705 ethnic groups recognized as Scheduled Tribes, which literally means Indigenous People. There are, however, many more ethnic groups that would qualify for Scheduled Tribe status, but which are not officially recognized. The largest concentrations of Indigenous Peoples are found in the seven states of North East India and the so-called central tribal belt stretching from Rajasthan to West Bengal. One of the characteristics of these indigenous people is that they live by riverbeds, forests, seashores and deep into hinterlands to protect their legacies. They depend on forest produce and locally available means of subsistence. As industries encroached upon their lands, many communities were displaced and some continued to wage a struggle to either protect their homes or demand fair compensation. By taking away forest lands for industries and plantation forestry instead of preserving natural species that provide livelihood to these people, the government has been depriving them of the basic means of livelihood.

Recognizing their rights to forest areas, and forest management practices, is critical to understanding their struggle for survival. Loss of forest cover, mining and the expansion of hybrid crops remain direct threats to the food security of these people who count on forest resources and wild food. There's a need for scientific discourse on the impact of climate change on species that grow in the wild and are used by indigenous people living close to forests.

The indigenous people of the North East, who are a complex mosaic of racial and cultural backgrounds, have retained ancient knowledge through oral storytelling, cosmological observations, cultural practices and rituals. Through literal and figurative narrative, music and dance, rituals and other means, this knowledge has been passed down through the ages. While animism is widely practised in the North East, there are significant differences in the deities, oral traditions, rituals, festivals, environmental ethics, sustainable agricultural methods and taboos about particular plants and animals in the region. Most indigenous people around the world have made living in peace with the environment an essential element of their cultures, traditions and lifestyles. Their customs, rituals and religious activities foster harmonious relationships between nature, the environment and the universe. Such a way of life provides insight into the indigenous people's cosmology, ontology and epistemology in addition to their knowledge systems. Ethnic and cultural diversity, as well as behaviours that preserve the environment and way of life, are all integral aspects of the North East.

The local communities have enormous collective moral and cultural capital as well as intangible cultural richness. The people view the forests, mountains and rivers as a storehouse of traditional knowledge and wisdom. Unfortunately, the so-called mainstream has relegated these people to the margins of conversations, reinforcing the myth of the helpless native.

The indigenous people have long maintained a way of life and cultural traditions that emphasize what social scientist Shiv Viswanathan refers to as "everydayness of coping." Their method of subsisting on the bare minimum is built around doing more with less. According to Vishwanathan, this "ethnography of coping" is a "lesson for us all." Their daily experiences give us a compass for understanding common knowledge and how the margins maintain and protect their knowledge systems. A sense of the collective has been a part of such a philosophy, worldview and lifestyle.

The tribal people of the North East Region had techniques for preserving, protecting and coexisting peacefully with nature. Analysis of the relationship between the environment and this region's sustainable development is urgently needed because the region's society is rapidly losing its last vestiges of happiness because of the exponential rate of change brought on by several problems that may or may not be directly related to the gene pool in the area. The best solution might involve combining conventional and contemporary methods and reviving ancient tribal knowledge and ideas about environmental protection.

In every developed and developing nation, there are indigenous communities and indigenous knowledge. The knowledge that is present in these societies is indicative of those who actively and intuitively interact with their natural and biological surroundings. While the interdependence of culture and nature is essential to these people, they exhibit diverse understanding and unique processes in environmental regulation. Their distinctive culture and heritage are so varied that they create characteristic traits for each tribe that are specific to their society.

In tribal communities in the North East, indigenous knowledge is crucial for making decisions about agriculture, healthcare, natural resource management and food preparation. In the Ziro Valley of Arunachal Pradesh, the Apatani eco-cultural landscape serves as a superb illustration of a district community-based natural resource management technique. This Arunachal Pradesh farming community is progressive and has its own local understanding of land management. In the Apatani valley, paddy-cum-fish cultivation occupies roughly 48.38% of the area, followed by clan woods (32.64%), bamboo forests (16.41%) and home gardens (2.75%). The Apatani tribe, which lives in the Ziro or Apatani valley, has been known to practise sedentary wet-rice cultivation integrated with pisciculture and rural forestry in the constrained flat valley land. This contrasts with the hilly regions of North East India, where shifting cultivation is the primary method of land use. In addition to maintaining sustained productivity levels, this traditional method is extremely energy efficient. The extensive traditional environmental knowledge system used by the Apatani clan to sustain their livable professions represents their position as skilled resource managers, which has also attracted UNESCO's attention to designate the Apatani/Ziro valley as a World Heritage Site.

The idea of a community existing in harmony with the environment is one of well-being, which is a culturally specific concept. The key concept that drives the definition of "well-being," in the worldviews of many indigenous peoples around the globe, is the harmonious coexistence of individuals with their local community and natural environs. True well-being comes from peaceful coexistence. Promoting people's livelihoods and the environment is its main goal; it is not income dependent.

Another small tribal group in Arunachal Pradesh is the Akas. They engage in shifting agriculture, or traditional jhum. The indigenous knowledge system of the Akas is helpful for preserving forest resources. Diverse viewpoints and acceptances are very important in protecting woodland resources, including sacred woods. The financial condition and manner of life of people are greatly impacted by backwoods and woodland products, and all human activities revolve around them. They worship the woodlands as *Thouw-new*, which means an attendant and a caretaker. They worship the mountains and waterways, and some of the areas are sacred forests. Such forests are rigorously shielded from intrusion and material extraction of any kind. Whatever the circumstance, they can still harvest materials for the construction of buildings and sustenance, using their ancient methods of hunting, fishing, food gathering, and extraction. They have been thrifty with abuse and protective of the environment due to the in-built traditional cultural and social instruments.

Nagaland, which has a wide variety of flora and animals, is one among the world's 18 mega biodiversity hotspots. Where the Arakan Patkai extension of

Myanmar meets the Eastern Himalayan range, the whole Naga region has hilly mountainous terrain, apart from sections in the South and West. Nagaland is a state with incredible socio-cultural diversity in addition to its perplexing physical aspects. It is home to 16 major tribes that represent several ethnic groups, each of which has its own culture, language, customs, legends and history. The remaining tribes all belong to the Naga group. Their rich folklore and folksongs reflect their profound knowledge of and reverence for nature and everything around them. Their intergenerational cultural evolution, survival, beliefs, spirituality and medical systems are all related to this. They have been worshipping nature and holding a firm belief that God, humans, animals, aquatic life and plants had coexisted together for all recorded time.

In terms of how the Nagas have been interacting with the natural world, observance of taboos and "genna" served as a guiding principle. Taboo denotes a prohibition on something that is thought to be improper for religious or other reasons, and the "genna" refers to refraining from travelling to the fields and working only on specific days. Taboos vary in length and severity depending on the relevant tribe and the situation. Observation of taboos might endure anywhere from a single day to several days or even weeks. It is also important to introspect and to consider their relationship with nature, agriculture, animals and the Supreme Being, which keeps them strong and resilient across varied circumstances.

Therefore, adhering to taboos is essential to preserving human-environmental ties and the interconnectedness of the Naga people with their environment. In a manner, the taboo has kept the Nagas from overusing nature while enhancing their spiritual comprehension of their physical surroundings. Therefore, it must be scrupulously adhered to to avoid effects on one's life that could result from fellow members of society or from supernatural entities.

One of the mega-biodiversity hotspots is the state of Meghalaya, which includes the Khasi, Garo and Jaintia hills. Along the mountain ranges of the Garo and Khasi hills areas, there are various sacred forests. The native tribes in the state oversee more than 90% of the state's woods. The people of Meghalaya consider the sacred forests to be the home of gods who protect them from evil spirits and bestow welfare upon them, their cattle and their land. These forests are home to a wide variety of orchid species, as well as medicinal and attractive plants, bamboo, canes, timber- and resin-producing plants and wild relatives of cultivated plant species. Most plants utilized in traditional medicine are those that are harvested from sacred woods. Only systematic removal of raw materials from the groves has been prohibited because of taboos, religious beliefs and folklore protecting therapeutic plants. Additionally, many species and birds that are endangered worldwide can be found in these sacred forests. The preservation of soil and water is greatly aided by them.

In upper Assam, the Tai-Khamyangs are regarded as one of the most knowledgeable clans in ethno-medicine. Their renowned traditional natural remedies are quite effective, and both Khamyang and non-Khamyang people from all around come to profit from the traditional specialists' administrations. Their vast knowledge of plant-based ethno-medicine and many magicostrict beliefs and practices are transmitted primarily orally and in some cases through Tai literary works from one generation to the next. The members of this network show a great appreciation for the surrounding plant resources. They are incredibly knowledgeable about the distinctive qualities of the environmental variables that form a fundamental part of both material and unearthly cultures. Part of the legendary Tai stock are the Tai-Khamyangs, also known as Khamjangs or Shyams. From South West China and North Myanmar, various Tai ethnic groups have historically relocated to Assam. After crossing the Patkai hills in the middle of the 18th century AD, the Khamyang people moved from North Myanmar to Assam.

Impact of Human Activities on Environment

The Anthropocene period has already begun, regardless of whether we are close to the "end of nature." It is the result of important planetary boundaries being violated by human activity. The effects of human activities on the environment, temperature and ecology of the Earth are so significant that they will leave a lasting mark in the stratigraphic record. The Anthropocene forces us to reconsider how we view the environment, life and our means of subsistence in addition to changing our relationship with the planet. In these unparalleled waves of adversity, only the lifestyles and cosmology of the indigenous people provide us several levels of resistance.

The indigenous people turn to their traditional ways to help them adjust to hardships, and resilience is anchored in traditional knowledge. Their resilience comes from maintaining connections with their families and communities. There may be poverty among the indigenous people, but there is no other issue because of their sharing culture and capacity to get more done with less. Even though life can be challenging, it is seen as an opportunity. The indigenous people rely on this resilience to deal with the climatic issue. This adaptability can be seen in the taboos on hunting and eating fish, waterfowl and other aquatic animals during specific times, as well as in the Meiteis of Manipur's gathering and consumption of specific flora and the preservation of sacred trees in Meghalaya.

It is therefore essential to critically examine current development trajectories and create new routes that result in outcomes that are sustainable from an economic, social and cultural perspective.

Cosmology of Conservation

The indigenous people of North East India have been practising their worldview and holding to a cosmology of conservation for thousands of years. In this cosmology, humans were one of many components, along with animals, birds, trees, clouds, mountains and earth. In this system, humans were not viewed as superior to other life forms or elements, and they shared responsibility for maintaining their environment. The organic cosmology has developed an ecological ethic that still directs the behaviour and practices of the indigenous people. Through social interactions, oral traditions, ritualistic activities, song, dance and other ceremonies, knowledge is transmitted formally or informally among kin groupings and the community.

The knowledge systems of each indigenous group in the North East are ingrained in their culture, history, relationships with the local economy, with people's lives and with how they perceive the world. However, their traditional wisdom is quickly disappearing. It is still mostly unreported, and given the speed of socioeconomic change, it could disappear or change significantly. This knowledge includes the complex collection of facts, insights and interpretations that direct human societies in their interactions with one another in a variety of contexts, such as agriculture and animal husbandry, fishing, dealing with diseases, describing natural phenomena and developing coping mechanisms for changing environments. The indigenous people and communities have left behind a solution that can benefit not only our societies but also the rest of the world.

Conclusion

People are becoming increasingly aware that involvement is essential for sustainable development on a global scale. The local and indigenous cultures' traditional knowledge appears to be an important strategy for the preservation and protection of biodiversity. The North East region's indigenous inhabitants seem to live in harmony with nature in a certain way. People in the North East region view life ethically. The tribal people live in community groups and are not materialistic. They care more about happiness than they do about material things. The symbiosis with nature is unequivocally demonstrated through research into the tribal culture and beliefs of this area.

Pragmatic efforts advocate for activities at all levels, including global, regional, national, local and community, to address environmental challenges. Control of natural resources will be greatly aided by the decartelization of power to the local level in the forest conservation policy. Giving communities legal rights and authority may be a useful strategy for the sustainable management of natural resources. The traditional expertise of managing natural resources will be a great justification for creating effective technology for the long-term conservation of biodiversity in the age of globalization. It is up to politicians and administrators to decide whether to combine traditional knowledge with modern technology for the benefit of the local population in general and the North Eastern region in particular. The North Eastern region's highly regarded century-old traditional ecological resource management is essential because eroding these principles of the traditional tribal community is causing disruptions in both social and ecological settings.

It is indicated that until and unless community involvement is made mandatory, no amount of policy making can bring about a collective change. The knowledge and the technical understanding that the indigenous people hold about their environment, the traditions that are practised for preservation, the customs that are upheld for conservation, are priceless. If they are incorporated and integrated with the government directives, there is a genuine potential for conservation, preservation and protection of the environment.

If humans are to survive, they must work together with *wild*; there is no other alternative.

Bibliography

Adger, W. N., Arnell, N. W., & Tompkins, E. L. (2014). Successful adaptation to climate change across scales. *Global Environmental Change*, *15*(2), 77–86. https://doi.org/10.1016/j.gloenvcha.2004.12.005

Assam History. (n.d.). *Assam State Portal*. https://assam.gov.in/about-us/394

Giz. (2024). *Climate change adaptation—North eastern region of India (CCA-NER)*. Startseite. https://www.giz.de/en/worldwide/16041.html

IPCC. (2014). Chapter 4: Water. Climate Change 2014: Impacts, Adaptation and Vulnerability. https://www.ipcc.ch/report/ar6/wg2/chapter/chapter-4/

Matt Murphy in London & Daniel Pardo in Bogota. (2023). *How children survived 40 days in Colombian jungle*. BBC News. https://www.bbc.com/news/world-europe-65871238

Mehta, V. (2023). India holds the key to hitting global climate change targets. *BusinessLine*. https://www.thehindubusinessline.com/opinion/india-holds-the-key-to-hitting-global-climate-change-targets/article66398563.ece

Ministry of Development of North Eastern Region, North East India, Government of India. (n.d.). *Ministry of development of North Eastern Region: Government of India*. https://mdoner.gov.in/

Ministry of Earth Sciences. (n.d.). *Home*. https://www.moes.gov.in/documents/annual-reports

Ministry of Environment and Forests (MoEF). (1990). Greenhouse effect and climate change: Issues for the developing countries. In *Proceedings of the Conference of Select Developing Countries on Global Environmental Issues*. MoEF, GoI.

MyGov Blogs. (n.d.). *How does India taken a leadership role in tackling climate change?* https://blog.mygov.in/editorial/how-does-india-taken-a-leadership-role-in-tackling-climate-change-3/

Pandve, H. T. (2009). India's national action plan on climate change. *Indian Journal of Occupational and Environmental Medicine*, *13*(1), 17–19. https://doi.org/10.4103/0019-5278.50718

Press Information Bureau. (n.d.). *India is part of the solution and is doing more than its fair share to address climate change*. https://pib.gov.in

Scheffran, J., Guo, W., Krampe, F., & Okpara, U. (2023). *Tipping cascades between conflict and cooperation in climate change*. https://doi.org/10.5194/egusphere-2023-1766

Sengupta, S. (2019a). India's engagement in Global Climate Negotiations from Rio to Paris. In *India in a warming world* (pp. 114–141). Oxford Academic. https://doi.org/10.1093/oso/9780199498734.003.0007

Sengupta, S. (2019b). *India's engagement in global climate negotiations from Rio to Paris*. OUP Academic. https://academic.oup.com/book/35227/chapter/299750865

Shaw, R., Luo, Y., Cheong, T. S., Abdul Halim, S., Chaturvedi, S., Hashizume, M., Insarov, G. E., Ishikawa, Y., Jafari, M., Kitoh, A., Pulhin, J., Singh, C., Vasant, K., & Zhang, Z. (2022). 2022: Asia Supplementary Material. In: Climate Change.

Impacts, adaptation and vulnerability. *Contribution of Working Group II to the Sixth Assessment Report of the Intergovernmental Panel on Climate Change* [H.-O. Pörtner, D. C. Roberts, M. Tignor, E. S. Poloczanska, K. Mintenbeck, A. Alegría, M. Craig, S. Langsdorf, S. Löschke, V. Möller, A. Okem, and B. Rama (eds.)]. https://www.ipcc.ch/report/ar6/wg2/

United Nations. (n.d.-a). *Climate change—United nations sustainable development.* United Nations. https://www.un.org/sustainabledevelopment/climate-change/

United Nations. (n.d.-b). *Climate change.* United Nations. https://www.un.org/en/climatechange

World Economic Forum. (n.d.). *India holds the key to hitting global climate change targets.* https://www.weforum.org/agenda/2023/01/india-holds-the-key-to-hitting-global-climate-change-targets-here-s-why/

Youdan, C., & Bajaj, P. (2022). *India's approach and position on India's climate change governance.* National Maritime Foundation. https://maritimeindia.org/indias-approach-and-position-on-climate-change-governance/

2 Climate Impact in North East India

Tsenbeni N Lotha, Ketiyala, Vevosa Nakro and Latonglila Jamir

The three biggest obstacles to sustainable development are seen to be climate change, global warming, and loss of biodiversity. Agriculture is undoubtedly the most climate sensitive sector due to its close relationship with climatic factors like temperature and precipitation. As a result, in recent years, the possible implications of climate change on this critical area of human welfare have received the most attention (Kumar et al., 2007). Every part of the world has already experienced the effects of human activity on climate including climatic extremes, like excessive heat waves, droughts, increased tropical activity, and rising sea levels (Pachauri and Meyer, 2014). Even though sceptics still maintain that all this change is occurring naturally and not because of human actions, science is clear that reducing greenhouse gas emissions will help mitigate the worst effects of changing climate (Kaplan and Ramanna, 2021).

Developing countries are more likely to experience negative effects from climate change because of their significant reliance on climate-sensitive livelihoods like rain-fed agriculture, water, and forestry (Moorhead, 2009). The International Red Cross Society affirms that there will never be a vaccination for climate change and that it is far more deadly for humanity than the COVID-19 pandemic (Walton and Aalst, 2020). The thickening layer of carbon dioxide pollution, mostly from power plants and vehicles, traps heat in the lower atmosphere and directly affects biodiversity, which in turn influences the decline in species variety and how well biological systems can sustain human demands (Saikia, Kumar, and Khan, 2016).

The North Eastern Region

Geographically North East India covers the seven sister states viz. Assam, Arunachal Pradesh, Manipur, Mizoram, Meghalaya, Nagaland, Tripura, and the eighth Sikkim. The area is situated roughly between longitudes 89.46 and 97.5E and latitudes 29° and 22°N. It covers an area of 262,185 square kilometres, or about 8% of the country's total land area. There are 389.84 lakh people living in the region with around 150 prominent indigenous communities, sheltered in humid evergreen, temperate, and alpine woodlands (Jain et al., 2013; Tynsong, Dkhar, and Tiwari, 2020). Nearly 55% of the overall

DOI: 10.4324/9781003504870-3

geographical area is covered in forest. China, Tibet, Bangladesh, Myanmar, and Bhutan are all located on each side of the states in the region. The territory has hilly terrain interspersed with valleys and plains where the altitude ranges from almost sea level to over 7,000 metres. This includes the Brahmaputra, Barak, and Imphal valleys as well as some flatlands between the hills of Meghalaya and Tripura (23,000 ft. above mean sea level). The region is known for its scenic landscape, cultural heritage, ethnic beauty, and richness of natural resources.

The tribes of the North Eastern region of India have distinctly different socio-economic and socio-cultural situations. The region is endowed with a thriving ichthyological field and a highly rich biodiversity (Subenthung Odyuo and Nagesh, 2012). Lowland tropical evergreen woods in this area are thought to represent the northernmost extension of true tropical rainforests on the whole planet (Proctor, Haridasan, and Smith, 1998).

Climate and Weather of North East India

North East (NE) India has the features of a tropical climate, especially in the lowlands, experiencing heavy rainfall throughout the four summer months of June to September. The region has three distinct seasons: winter, summer, and the rainy season. The mountainous areas and the valleys have different climates (Varah and Varah, 2022).

A network of weather stations equipped with weather instruments and even transmitters records weather conditions on a daily or even hourly basis in the NE region. More than 36 weather stations have been set up in different regions of the NE states. These weather stations directly transmit the weather data to the Central Meteorological Bureau for compilation and public dissemination. The density of weather stations, which also defines the level of prediction accuracy, creates a generalised image of the climate. Temperature, humidity, and rainfall conditions in a location are frequently calculated over a 30-year period and expressed in terms of averages (Mahanta et al., 2021; Yadava, Yadav, Misra, Singh, and Singh, 2015).

The country's wettest region is NE India where the highest annual rainfall in the world, on the Meghalaya plateau, is recorded in Cherrapunji (25°–15′ N, 91°–44′ E), 50 km south of Shillong. The southern and western borders of the region get an early monsoon burst in May or June, with June having the most rainfall. Arunachal Pradesh, Nagaland, Manipur, Mizoram, and Meghalaya are among the hilly and mountainous regions, where temperatures are often lower than those in the plains throughout the year. In the plains, the summertime temperature ranges from 30 to 33°C while the Western section is at their peak in April. The greatest temperatures of 34.2 and 32.7°C are recorded in April in Agartala, Tripura, and Dhubri in West Assam.

Causes of Climate Change in North East India

In NE India, there have been some observable changes in temperature and rainfall patterns over the past century, indicating that climate change is rapidly becoming an unpleasant reality of the present rather than a hypothetical future concern. Recent studies have demonstrated that anthropogenic forces brought on by changes in land use and land cover (LULC) is one of the drivers of the recent temperature shift. These changes in LULC influence the occurrence of heat or cold waves, clouds, and rainfall patterns. Furthermore, LULC change has been associated with atmospheric aerosol emissions, which can alter the surface temperature through both direct and indirect effects, hence regulating rainfall, which can result in droughts or floods given certain favourable conditions (Gogoi et al., 2019; Stocker et al., 2013).

A variety of terrain features, monsoonal currents during the South West and North East monsoon seasons and other natural elements all contribute to NE India's subtropical climate. The massive Himalayas in the north, the plateau of Meghalaya in the south, and the hilly regions of Manipur, Mizoram, and Nagaland in the east all have an impact on the climate because of the region's distinctive geographic location. Additionally, the Bay of Bengal's closeness to the region affects the climate (Rampini, 2016). Overall, the territory's peculiar topography and unique geographic location results in a complicated interaction between the land, sea, and atmosphere that causes the region to have certain climatic characteristics (Dimri, Kumar, and Srivastava, 2018).

Vulnerability to Climate Change

The general definition of vulnerability is "the potential for harm in the form of injury, death, loss, interruption of livelihood, or other harm in the event of an extreme occurrence, or unusual difficulties in recovering from such consequences" (Wisner, 2002). Ravindranath and others (2011) used the outcome vulnerability assessment and developed vulnerability indices to understand the vulnerability of three key sectors of the NE region. NE India is already incredibly sensitive to the effects of climate change because of its low levels of economic development, relative political and economic marginalisation, and the fact that local businesses and livelihoods depend significantly on the Brahmaputra River and riparian habitats. There is a pressing need to comprehend how 140 new dams in the river basins will affect, and how the vulnerable local people will adapt to, the changing water resources as a result of climate change (Rampini, 2016).

Impact of Climate Change in North East India

One of the greatest environmental threats the world is facing today is climate change, which has the potential to have an adverse effect on human health, food production, water security, biodiversity in forests and other natural ecosystems, and human settlements (Ravindranath et al., 2011). The variation in the natural environment is brought on by the climatic circumstances, which modify the planet fundamentally Nayak and Tekemi, 2022). NE India, where these variations are significantly more apparent than in other parts of the country, is beginning to show signs of climate change. Different trophic levels are affected by climate change due to a variety of factors, including temperature, rainfall, CO_2 concentration, and extreme events like droughts, floods, and fires. These climatic forces can manifest at varying intensities and may have an impact on the living systems at the population, species, and ecological levels (Gates, 1990).

Impact on Agriculture and Food Security

Due to the close relationship between agricultural and climatic variables, the influence of climate change on agriculture and food security has been at the heart of research and policy objectives (Arias et al., 2021). Though NE India's agriculture is regarded to be the most climate-sensitive in the country, little is known about how climate change may impact food production (Kumar et al., 2007). Dipak Sharma and co-workers (2021), in their recent study, have reported the variability of climate in the state of Arunachal Pradesh which poses a threat to agriculture and food security and have stressed on the importance of climate-resilient agriculture ensuring food security in the present and future. Because of the increasing temperatures and erratic rainfall, some types of vegetables and edible ferns that are essential to Assamese cuisine are now going extinct (Rahman, 2016).

A study conducted by Nayak and co-workers (2021) to study the influence of LULC on climate change over North India and NE India (NEI) during the period of 1981–2006 used OMR (observation minus reanalysis), which was accomplished by subtracting the reanalysis temperature from the observed temperature. According to their study, it was found that North India had a warming of 0.03°C while NE India had a cooling of temperature of about 1.5°C during the period of 1981–2006. The extension of the dry land and the reduction of the dense forest were found to be the causes of the warming over North India between 1981 and 2006, while on the other hand, the decrease in non-vegetated areas and the growth of agricultural land covers throughout that time are responsible for the cooling over NE India. The reason for the decreased areas under these categories over NEI could be attributed to the snowmelt and an expansion of monoculture farming. The cooling over NEI could also be associated with the large increase of dense forest. The areas under dense forest over NEI were increased by 0.3% during 1981–1990 and again increased by 0.5% during 1991–2006; increase in dense vegetation land cover controls the temperature over a region more effectively than the land with less vegetation cover (Nayak et al., 2021).

According to the Indian Meteorological Department's (IMD) (2021) report (IMD, 2021), Arunachal Pradesh has experienced severe deficiency in monsoon rains which shows a strong likelihood of extremes in temperature and rainfall, especially during the monsoon season, causing serious harm to agricultural crops. These local varieties of crops are highly vulnerable to additional negative consequences of climate change, such as changes in pest and disease activity, changes in soil temperature and moisture, and changes in soil mineral makeup. Due to low productivity, the issue of food insecurity may eventually arise and affect the most susceptible of society, that is, the poorest members in farming communities (Viswanathan, Kavya, and Bahinipati, 2020).

Impact on Economy and Society

The economy of the North Eastern region is mostly rural, underdeveloped, and localised, favouring Jhum agriculture. The region has a large number of streams, lakes, and rivers and other natural freshwater bodies, making it an area with the most freshwater resources. Due to the mountainous geography and various other regional logistical problems, industrialisation has proven difficult. As such, of significant importance are the ecosystem services that forest and other components of this area provide.

The alpine ecosystem is extremely vulnerable to changes in the climatic pattern due to its geo-ecological fragility, placement in the Himalayan topography, transboundary hydrological regimes, and inherent socio-economic instabilities. In order to provide reliable climate projections for future carbon-intensive scenarios that are consistent with growth, the past performance of the climatic regime must be investigated. This will help with developing and putting into practice relevant adaptation and mitigation strategies in addition to helping with the evaluation of past changes in the study region (Dimri et al., 2018).

The loss of life, property, livestock, and crops has occurred as a result of landslides that are often a common concern in hilly areas such as NE India. The health service, including water, electricity, and communication lines, is shown to be greatly impacted (Ramesh and Girish, 2022). For instance, in a research study conducted by Pentile Thong and co-workers (2022) on the susceptibility of tribal households in Mizoram to climate variability, using the lens of a socio-economic vulnerability framework, confirms that the Jhumia households who are basically engaged in Jhum fall short of implementing any adaptation strategies to the impacts of climate change and are most vulnerable. Additionally, it was found that Muallungthu village from the state of Mizoram was the most vulnerable village socio-economically. The rising temperatures and decreasing precipitation in rain-fed agriculture can shorten the growing season and decrease crop productivity, which influences the economy of the farmers.

Impact on Biodiversity

Biodiversity loss is defined by the Convention on Biological Diversity (Stock, 1992) as the long-term decline in components of biodiversity and their capacity to create goods and services. Studies show that the NE region of India experienced the greatest decline in tree cover in 2020, losing 79% of its total area under trees. Almost 110,000 hectares of tree cover were lost in the region, according to statistics on forest change collected by the University of Maryland (Hazarika and Bhattacharjee, 2022). The NE region jointly lost 5–14% of the country's forest area between 2001 and 2020, while Nagaland experienced its first forest cover decline by 17% in the last two decades, which is the fastest since 2001. Assam, the largest of the seven sister states, contributed to 14.1% national tree cover loss during the period. Tripura came in second with a decrease of 15% (Roy, 2021).

Impact on Human Health and Livelihood

The global increase and intensification of rainfall extremes have been noticed over the past ten years, and new evidence indicates anthropogenic climate change as the cause, but more quantification is required to completely grasp the human influences on severe occurrences (IPCC, 2021). Extreme rainfall over NE India was a catastrophic but little-studied event. This extended period of excessive precipitation, known as the 2020 event, resulted in numerous hydrogeological catastrophes throughout NE India and other South Asian nations. According to reports, this terrible catastrophe left 9.6 million people affected, 550 individuals dead or missing, and a direct economic loss of up to hundreds of millions of dollars (IFRC, 2020).

Studies project that because of climate change, rural people have relocated in search of better opportunities for a living. Due to factors including increased warmth and dryness in some locations, there is a limit in agricultural production, and increased precipitation, which may cause flooding in low-lying areas (Shamsuddoha and Choudhury, 2009). By 2030, the temperature in Nagaland is expected to rise by 1.6–1.8°C. With rising temperatures, there may be an increase in dehydration among the population, as well as an increase in skin diseases such as acne, cellulitis, and furunculosis, particularly in the southern districts. From 2008 to 2010, Nagaland's acute respiratory disease rate nearly tripled. In Dimapur, the respirable suspended particulate matter (RSPM) level has risen above the allowed national standards of 60 g/m^3 and 14,060 g/m^3 due to pollution (Yepthomi, 2016).

Impact on Fisheries

The region's temperature and rainfall patterns have considerably changed. The annual average temperatures in and around Guwahati, Assam, exhibited an increasing tendency during the past 37 years (1980–2017), bringing absolute changes with mean maximum and mean minimum temperature of 2.6 and 2.2°C. The annual average rainfall, on the other hand, exhibited a downward tendency for the same area and time period, with an absolute change of 935.4 mm. But every year, particularly in Assam, floods wreak havoc on the region, destroying fisheries, crops, animals, and people's lives and livelihoods. It is anticipated that both the intensity of floods and the severity of droughts would increase in many regions of NE India (Ravindranath et al., 2011; Yengkokpam et al., 2022). Forecasts of how climate change will affect inland fisheries must be considered by managers and fishers in response to these changes, which could either worsen or reduce the consequences of climate change on the aquatic biodiversity and eventually on the fish consumers. For example, it is apparent that flood-related chaos has a significant influence on the lives and livelihoods of fishers, as well as on the socio-economic conditions of fish breeders and other stakeholders of NE India. Therefore, if harvest rates are not decreased in response to diminishing productivity, the risk of overfishing may increase, which might then weaken the capacity of stocks to tolerate climatic change and affect long-term output (S.K Das, 2020).

Climate Crisis in NE India

North East India's climate is changing, and over the past century, the region's rainfall patterns have seen significant change. The drying of the land, which raises the frequency and severity of dry periods and droughts, is a warming-related factor that affects rainfall. The flow of rivers, the amount of snow cover, and the health of mountain springs are all impacted by the changing rainfall pattern, especially during the monsoon season. These factors in turn have an effect on livelihoods, particularly agriculture and fishing, forest flora growth, animal and bird habitat, and other ecosystem aspects.

Some events indicative of climate change from the various states of NE India are given in Table 2.1.

Table 2.1 Climate crisis in NE region of India

State	*Description*
Assam	Assam's climate is distinguished by high humidity. Its most notable feature is the considerable rainfall between March and May, when precipitation in upper India is at its lowest (Upadhyay et al., 2009. Furthermore, fluctuations in climatic conditions like temperature and rainfall, and increase in the frequency of extreme occurrences such as flood and drought, have wreaked havoc in recent years (Buragohain, Hazarika, and Barman, 2019). The abrupt and steep transition from the mountains to the plains makes the region naturally prone to flooding. Increased rainfall causes tributaries to flood more frequently, increasing the likelihood of inundations and droughts in these floodplains and those to the south of Brahmaputra, where rainfall trends have been declining (Begum and Mahanta, 2017; Borah et al., 2022). The plains of the Brahmaputra valley regularly accommodate monsoon-driven flooding of the Brahmaputra River and its tributaries, causing widespread flood inundation and geomorphic changes. Since Assam floods are known to be among the worst disasters in recorded human history, a synthetic aperture radar–based approach has been used for a thorough flood inundation mapping of the region (Devrani, Srivastava, Kumar, and Kasana, 2022).

(*Continued*)

Table 2.1 (Continued)

State	*Description*
Arunachal Pradesh	The area is largely mountainous, with Himalayan ranges running along the northern boundaries and mountain ranges that stretch from snow-capped mountains in the north to the plains of Brahmaputra valley in the south (Thungon, 2020). Climate change has altered the patterns of snowfall and rainfall in higher elevation zones. Forest fires, landslides, mud/debris flows, earthquakes, cloudbursts, and floods have had an impact in the mountainous region at different times and in different places (Purabi Saikia, Kumar, Lal, and Khan, 2020). In 2020, there was an increase in rainfall which led to the decline in rice productivity. Increased soil and atmospheric moisture have led to the emergence of new illnesses, as well as an increase in pest infestations. The region's agricultural productivity has greatly decreased as a result of rising temperatures and increased rainfall (Sangomla, 2021).
Manipur	Manipur continues to experience the negative effects of climate change on human security and are plagued by destructive storms, continuous rain, landslides, fissures, and habitation being washed away by rivers that have overflowed their banks (Salam, 2017). The valley's population is approximately 67% farmworkers, exerting tremendous pressure on a tiny area to feed its entire population (Laitonjam and Feroze, 2017). Only 10% of Manipur is a valley where a little portion of it is used for agriculture as one-fourth of the area is taken up by lakes, swamps, dry land, and small hillocks. Shifting rainfall patterns have influenced agriculture, which has caused impact on food security (http://desmanipur.gov.in). The National Innovation on Climate Resilient Agriculture (NICRA) aims at strategic research on adaptation and mitigation, demonstration of technologies on farmers' fields, and creating awareness among farmers and other stakeholders to minimise the climatic change impacts on agriculture. Trend analysis of weather variables in Imphal under NICRA revealed that the mean annual maximum temperature has been increasing (0.10°C per decade). In 2019, a staggering 56% rain deficit was also reported, and season 2021 was not any better. The state's declining rainfall trends have had an impact on agriculture productivity, particularly paddy and fish farming (Bora, Bhuyan, Hazarika, Gogoi, and Goswami, 2022; Singh et al., 2011).
Meghalaya	Meghalaya experiences the heaviest rainfall in the world (Khanum, 2021), where 71% of its annual rainfall arrives during the monsoon season (Rao et al., 2011). Due to its location in seismic Zone V, it carries a high possibility of earthquakes where Khasi hills region is extremely susceptible to them. In addition to earthquakes, there is risk for other weather-related disasters such as flash floods, landslides, strong winds, and thunderstorms (Magotra et al., 2021).

(*Continued*)

Table 2.1 (Continued)

State	*Description*
	In 2019, widespread landslides were caused by persistent rain. The Umkhrah River in Shillong City reached dangerous water levels (TNN, 2018. Between the 30-year period from 1909 to 2018, the state's total rainfall trend decreased, and it became more pronounced starting in 2005 (Marak, Sarma, and Bhattacharjya, 2020). Shillong experienced more than 72.2 mm of rain in July 2019, which caused low-lying portions of the city to be flooded. According to the Water and Climate Lab at the Indian Institute of Technology, the state's ambient air temperature is rising at a rate of 0.031 degrees per year and is projected to reach 1.6–1.7°C by 2050 (Dinish C Sharma, 2017a).
Mizoram	Mizoram experiences strong rains from May to September with an annual average rainfall of 245 cm due to the direct influence of the monsoon (Rangnekar, 2006). It is one of the areas in the Eastern Himalayan region with a high forest cover but high rainfall, extensive soil erosion, and floods. From 1951 to 2017, Mizoram experienced a yearly temperature increase of 0.01°C, which was primarily caused by an increase in the mean maximum temperature of 0.03°C per year (Sahoo et al., 2018). In 2019, three buildings constructed under the Basic Service to Urban Poor (BSUP) scheme at Durtlang collapsed due to landslide resulting in three confirmed deaths and eight others injured. This natural disaster brought on by climate change resulted in severe rain, blocked roads, and falling debris (Hmingthangzuala, 2019). A crucial impact is played by the slope, weathering, erosion, vegetation cover, and undercuts. Landslides are also known to be triggered by cloud bursts, floods, mining, dam collapse, building constructions, roads, or other land cover changes (Mishra et al., 2021).
Tripura	In the 45 years between 1953 and 2004, seven North Eastern states, including Tripura, together lost roughly 2 billion rupees because of flood damage to public infrastructure, including 1.25 million hectares of land and crops. As a result of the flood, Tripura experienced erosion on about 108,000 hectares (Das et al., 2009). The rainfall in Tripura varies significantly throughout the year. The particular problem is that all districts now have lower rainfall overall during the monsoon season, which makes up 60% of its annual rainfall of the past. There is a decreasing trend in the seasonal and annual average rainfall at most stations for the period 1986–2019.The monsoon season brings the majority of Tripura's annual rainfall. Winter and the post-monsoon seasons have the least rain (Murasingh et al., 2022).

Mitigation and Adaptation to Climate Change

The population of the North Eastern region relies on agriculture, which in turn depends heavily on climate suitability. Even a slight unfavourable change in the climate could potentially impede the people's ability to provide for their basic needs (Kumar et al., 2007). Many different adaptations have been suggested for the conservation of biodiversity, including assisted colonisation, using invasive species to restore ecosystem functions, and building and expanding protected areas, networks, and resource efficiency. The major components of sustainable development are believed to be the provision of affordable energy, management of water systems in support of agriculture, industry, human settlements, increased agricultural production, and preservation of soil, surface, and groundwater quality, thus impeding further loss of biodiversity.

Mitigating climate change involves both government effort and public support. India's North East Climate Change Adaptation Programme (NECCAP) is one of the first attempts worldwide to assess future trends for a large region with scientific methods in order to draft and implement specific adaptation measures. The purpose of the programme is to strengthen the adaptive capacities of target groups, thereby reducing their vulnerability to climate change in a target-oriented manner. This is to contribute to lowering the risks of climate change and to contribute to the adaptation to climate change in the North East region of India.

Indigenous people in North East India have developed a system of traditional knowledge over many decades and centuries that has taught them how to mitigate the effects of climate change. Traditional farmers often adapt to climate change by choosing crop varieties and management practices like choosing drought-tolerant local varieties, ensuring adequate water supplies, and planting plant species that can increase carbon sequestration. Other traditional farming systems and food production techniques include extensive crop planting, mixed cropping or agroforestry practices, opportunistic weeding, wild plant gathering, and others. These people are exceptional observers and interpreters of environmental change due to their collective knowledge of the land, sky, and sea. They are also capable of predicting the future evolution of climate variables. Additionally, these tried-and-true knowledge-based adaptation and mitigation measures maintain the resilience of social-ecological systems in a vulnerable ecosystem inhabited by traditional groups (Shimrah, 2017).

Climate-Smart Agriculture

Climate-smart agriculture is a technique for reforming and reorienting agricultural systems to ensure food security considering the new realities of climate change. In areas that are severely impacted by climate change, developing climate-resilient agriculture can help to ensure food security and an income for rural people. Communities that are vulnerable to climatic hazards should be able to anticipate risks so that they can react in an emergency (Taylor, 2018). For instance, potato cultivation in Meghalaya has shown to be negatively impacted by climate change, so farm-level adaptation strategies have been employed. The main barriers to the implementation of adaptation practices are the scarcity of high-quality seeds, market price volatility, lack of technical expertise regarding improved potato cultivation, ineffective transportation, marketing infrastructure, unavailability of production inputs, and lack of knowledge regarding climate change. Integrating local farmer adaptation methods into scientific research is critical for developing effective, farmer-friendly, cost-effective, and long-term adaptation techniques (Kharumnuid, Rao, Sudharani, and Kumar, 2017). Crop rotation, crop diversification, the use of xerophytes crop varieties, the preservation and use of local genetic varieties, soil organic management, water harvesting, and other practices like modern irrigation techniques for combating deficit rainfall by using the drip irrigation method and sprinkler irrigation techniques are some management strategies for climate-resilient agriculture (Molotoks, Smith, and Dawson, 2021).

NICRA: National Innovations on Climate Resilient Agriculture

NICRA is a smart and noble concept established by Indian Council of Agricultural Research, with its main goal to increase the adaptability of Indian agriculture, including crops, livestock, fisheries, and climatic fluctuations. The megaproject's three main objectives are strategic research, technology demonstrations, and capacity building (Lenka, Panigrahi, Mohapatra, and Patnaik, 2022). Over one lakh farm households from all over the nation are involved in the programme, which is being implemented in 100 districts. In 2011, this project was carried out in Krishi Vigyan Kendra at Ri Bhoi district of Meghalaya. Since its inception, the NICRA project has promoted and supported the adoption of a variety of technologies and improved methods for reducing climate change in order to encourage farming as a business, particularly in the hilly region (Neog et al., 2018).

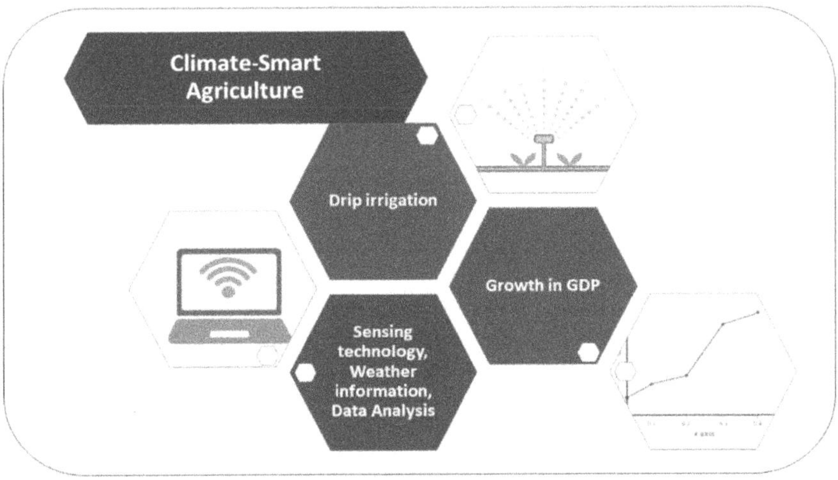

Figure 2.1 Objectives of climate-smart agriculture.

Traditional Strategies to Mitigate Climate Change in North East India

Traditional knowledge is the accumulated inherited understanding of farmers' environments, which include land, water, trees, plants, and other living things. Farmers have used this knowledge to solve problems by making a number of decisions and putting those decisions into practice by effectively allocating resources. It is utilised in North East India for a variety of purposes, including weather forecasting, promoting healthy seed germination, managing soil and water fertility, controlling insect pests, animal, plant illnesses, processing, storing, and fishing (De, 2021).

Traditional adaptation techniques such as bamboo drip irrigation, home gardens, and wet rice agriculture are common in NE India. In Nagaland, traditional coping mechanisms have been used and researched with regard to handling drought-like conditions. The production of food in Nagaland is being impacted by increasing temperatures, irregular rainfall, and the frequency of mild drought. To tackle this issue, the state is now turning to untapped resources and traditionally grown crops which can survive increased temperature and water pressure. In the wake of climate change, farmers will be encouraged to choose these types since they are the most likely to adapt (Dinesh C Sharma, 2017b).

People of Assam use traditional flood mitigation methods. *Kasa pithiya*, or traditional man-made mounds, are used to solve the problem of waterlogging. These mounds are typically built in the shape of a semicircle, resembling a turtle's shell rising above floodwater (Hussain et al., 2022). The Mishings who live in the valley's active floodplain build *Chaang ghor* to cope with flood water and to avoid earthquake damage (Bhuyan et al., 2022). In the absence of a credible flood alert system, farmers prepare for natural calamities using traditional knowledge. For instance, when ants move to higher ground, it predicts severe rain, alerting residents to a flood (Bhuyan et al., 2022; Mago et al., 2022). Returning to traditional rice varieties has proven beneficial to farmers in Assam who have faced multiple floods and drought-like conditions with equal severity. Flash floods during the monsoon season can last seven to fifteen days and are expected to worsen in the future due to climate change, so in order to prevent this, traditional rice varieties can be planted a little later than usual as it can even endure a drowned field (Rahman et al., 2016).

Indigenous people in Manipur are rural dwellers who continue to practise customary biodiversity preservation practices passed down from their forefathers, including botanical folklore. Plants are used not only for food, shelter, and clothing, but for learning and understanding how to predict the weather, foretell natural disasters, alleviate ill fortune, and, most importantly, in traditional healing practices too Singh et al. (2011). Bench terrace agriculture practised in the district of Chandel, Manipur, has proven to be an effective

Figure 2.2 *Chaang ghor* constructed by the Mishing tribe in the state of Assam to cope with flood water.

technology for preventing soil erosion in the district's hilly sloping areas (Singh and Kendra, 2021).

An ingenious method of employing bamboo pipes to tap spring and stream water to irrigate plantations is widely used in Meghalaya. It is so well developed that 18 to 20 litres of water that enter the bamboo pipe system each minute are reduced to 20 to 80 drops per minute at the plant site after travelling for several hundred metres. Typically, the betel leaf or black pepper crops grown in mixed orchards are watered using a bamboo drip irrigation system (Devi, 2018). Tripura farmers use organic manure to adapt to climate change. For instance, many farmers utilised cow dung by combining it with other organic resources, such as grasses, fallen leaves, and small tree branches, thus allowing them to decompose for growing summer potato crops (Dam, Das, and Das, 2022).

The majority of people in NE India practise ethnomedicine to treat a variety of illnesses. Ethnozoological methods are an important part of traditional treatment for people in Arunachal Pradesh, Nagaland, and Mizoram. However, due to various socio-political and economic changes in the way zoological medicines are harvested, many biological resources are on the verge of extinction. As a result, some ethnomedicinal practices must be improved and validated in order to reduce pressure on animal populations and sustain bioresources from diverse ecosystems. Education can help to generate a better understanding of how to support their conservation through integrated strategies and practices. This will aid in overcoming the problems caused by climate change and economic globalization (Chinlampianga, Singh, and Shukla, 2013).

Conclusion

India's North East is abundant like no other region in India. The climate change phenomenon has wreaked havoc as evidenced by the massive changes in weather conditions in the region. NE India can be considered as a hotspot for replenishing the declining biodiversity and changing climatic conditions. To achieve this goal, the states must learn to adapt by taking emerging measures as well as falling back on strategies that are already in practice. Effective mitigation methods for a climate-resilient agriculture production system in NE India must be developed and promoted along with risk management and support to farmers to increase resilience, which is the key to a successful change adaptation.

Going back to move forward with adaptation and resilience is perhaps the most important step that the people of the North East can take to mitigate the effects of climate change. By going back to the traditional practices in all spheres of their lives, including agriculture, fisheries, medicine, irrigation, shelter, and so on, perhaps the people will be enabled to counter their vulnerability to climate change.

References

Arias, P., Bellouin, N., Coppola, E., Jones, R., Krinner, G., Marotzke, J., & Naik, V. (2021). The Physical Science Basis. *Contribution of Working Group 14 I to the Sixth Assessment Report of the Intergovernmental Panel on Climate Change.* Technical Summary.

Begum, A., & Mahanta, R. (2017) Adaptation to climate change and factors affecting it in Assam. *Indian Journal of Agricultural Economics*, 72(3):446–455.

Bhuyan, G. V., Pradhan, D., Das, B. R., Bhuyan, M. J., Saha, S., Deka, N., Bhattachaya, P., & Bhagabati, A. K. (2022). Community response to local environment: Reflections in certain micro-landscapes of the Brahmaputra River Valley in Assam, India. *Journal of Kagawa University International Office*, 14(3), 278–286.

Bora, S. L., Bhuyan, L., Hazarika, P. J., Gogoi, J., & Goswami, K. (2022). Analysis of rainfall trend using non-parametric methods and innovative trend analysis during 1901–2020 in seven states of North East India. *Current Science*, *122*(7), 801–811. https://doi.org/10.18520/cs/v122/i7/801-811

Borah, L., Kalita, B., Boro, P., Kulnu, A. S., & Hazarika, N. (2022). Climate change impacts on socio-hydrological spaces of the Brahmaputra floodplain in Assam, Northeast India: A review. *Frontiers in Water*, *4*. https://doi.org/10.3389/frwa.2022.913840

Buragohain, D., Hazarika, U., & Barman, U. (2019). Climate variables, rice yield and farmers perc eption—A study in Nalbari District of Assam. *Indian Journal of Extension Education*, *55*(1), 109–114.

Chinlampianga, M., Singh, R. K., & Shukla, A. C. (2013). Ethno zoological diversity of North-east India: Empirical learning with traditional knowledge holders of Mizoram and Arunachal Pradesh. *Indian Journal of Traditional Knowledge*, *12*(1), 18–30.

Dam, S., Das, R., & Das, S. R. (2022). *Prospect of cow dung as a source of renewable energy in Tripura, India: Viability and challenges.* Retrieved from 10.4018/978-1-6684-4610-2.ch008

Das, A., Ghosh, P. K., Choudhury, B. U., Patel, D. P., Munda, G. C., Ngachan, S. V., & Chowdhury, P. (2009). Climate change in North East India: Recent facts and events – Worry for agricultural management. *Proceedings of the Workshop on Impact of Climate Change on Agriculture*, *2009*(December), 32–37.

Das, S. K. (2020). *Impact of climate variability on aquaculture and fisheries of North East India*. Indian Council of Agricultural research.

De, L. (2021). Traditional knowledge practices of North East India for sustainable agriculture. *Journal of Pharmacogmosy and Phytochemistry*, *10*(1), 549–556. Retrieved from www.phytojournal.com

Devi, C. V. (2018). Participatory management of irrigation system in North Eastern Region of India. *International Journal of Rural Management*, *14*(1), 69–79. https://doi.org/10.1177/0973005218765552

Devrani, R., Srivastava, P., Kumar, R., & Kasana, P. (2022). Characterization and assessment of flood inundated areas of lower Brahmaputra river basin using multi-temporal Synthetic Aperture Radar data. *Geological Journal*, *57*(2), 622–646.

Dimri, A. P., Kumar, D., & Srivastava, M. (2018). Regional climate changes over North East India: Present and future. In *Development and disaster management* (pp. 41–63). Springer. https://doi.org/10.1007/978-981-10-8485-0_3

Gates, D. M. (1990). Climate change and forests. *Tree Physiology*, *7*, 1–5

Gogoi, P. P., Vinoj, V., Swain, D., Roberts, G., Dash, J., & Tripathy, S. (2019). Land use and land cover change effect on surface temperature over Eastern India. *Scientific Reports*, *9*(1), 1–10. https://doi.org/10.1038/s41598-019-45213-z

Hazarika, B., & Bhattacharjee, N. (2022). Population pressure and its impact on forest resources in North East India. *Journal of Positive School Psychology*, *6*(4), 4245–4255.

Hmingthangzuala. (2019). *Landslide at Durtlang BSUP building complex, 3 confirmed dead*. Directorate Of Information & Public Relations, Government of Mizoram, India. https://dipr.mizoram.gov.in/post/landslide-at-durtlang-bsup-building-complex-3-confirmed-dead

Hussain, M. M., Bhuvaneswari, G., Borah, R. R. (2022). Transitioning to sponge city: A key to water resource management during floods in Assam. *ECS Transactions*, *107*(1), 4999.

IFRC. (2020). Homepage. *World disasters report 2020: IFRC*. https://www.ifrc.org/document/world-disasters-report-2020

IMD. (2021). India Meteorological Department (Ministry of Earth Science). https://Mausam.Imd.Gov.in/Imd_latest/Contents/Ar2021.Pdf. Retrieved from https://mausamjournal.imd.gov.in/

IPCC. (2021). Climate Change 2021: The Physical Science Basis. Working Group I Contribution to the IPCC Sixth Assessment Report. *Climate Change 2021: The Physical Science Basis*, 3949. https://doi.org/10.1017/9781009157896

Kaplan, R. S., & Ramanna, K. (2021). Accounting for climate change. *Harvard Business Review*, *99*(6). https://doi.org/10.2307/j.ctvz0h891.7

Khanum, F. (2021). Climate change: A threat to national security. *International Journal of Multidisciplinary Education Research*, *10*(10), 84–88.

Kharumnuid, P., Rao, I. S., Sudharani, V., & Kumar, S. (2017). Farm level adaptation practices of potato growing farmers in East Khasi Hills district of Meghalaya, India. *Journal of Environmental Biology*, *39*(9), 575–580.

Kumar, A., Valecha, N., Jain, T., & Dash, A. P. (2007). Burden of malaria in India: Retrospective and prospective view. *The American Journal of Tropical Medicine and Hygiene*, *77*(6 Suppl), 69–78.

Laitonjam, N., & Feroze, S. M. (2017). *Meteorological drought and Farmers' Perception of climate change: A case study of Manipur*. Sustainability.

Lenka, S., Panigrahi, R. S., Mohapatra, B., & Patnaik, D. B. R. (2022). Knowledge level of farmers on climate resilient agro-technologies curtail the climatic change effect in the NICRA district of Odisha. *The Pharma Innovation*, *11*(4), 13–17.

Mago, P., Sundaram, S., Singh, A., Gosain, N., Sarin, L., Mishra, S., Sharma, B., Bhagat, M., Kumar, T., Anand, R., Singh, R., Manda, S., Sirohi, S., & Roy, S. (2022). Sustainable environmental and cultural practices of Majuli, India: A case study. *SSRN Electronic Journal*, (06). https://doi.org/10.2139/ssrn.4148279

Magotra, R., Shaw, M., Parikh, J., Tyagi, A., & Sharma, Y. (2021). Disaster vulnerability assessment of Shillong, Meghalaya. *Urban India*, *41*(1), 147–165. Retrieved from https://irade.org/DisasterVulnerabilityAssessment.pdf

Mahanta, R., Saha, P., Rajesh, P. V., Nandy, S., Zahan, Y., & Mahanta, A. (2021). Reconstruction of a long reliable daily rainfall dataset for the North East India (NEI) for extreme rainfall studies. *Journal of Earth Science & Climatic Change*, *12*(9), 1–14.

Marak, J. D. K., Sarma, A. K., & Bhattacharjya, R. K. (2020). Innovative trend analysis of spatial and temporal rainfall variations in Umiam and Umtru watersheds in Meghalaya, India. *Theoretical and Applied Climatology*, *142*(3–4), 1397–1412. https://doi.org/10.1007/s00704-020-03383-1

Mishra, S. P., Puhan, J. P., & Mishra, S. (2021). Probing the crisis of regional connectivity instigated by the natural disasters, Mizoram, India. *International Journal of Environment and Climate Change*, *11*, 39–59.

Molotoks, A., Smith, P., & Dawson, T. P. (2021). Impacts of land use, population, and climate change on global food security. *Food and Energy Security*, *10*(1), 1–20. https://doi.org/10.1002/fes3.261

Moorhead, A. 2009. *Climate, agriculture and food security: A strategy for change*. Alliance of the CGIAR Centers.

Murasingh, S., Kuttippurath, J., Raj, S., Jha, M. K., Varikoden, H., & Debnath, S. (2022). Trends and variability of rainfall in Tripura State of India in 1986–2019 and key drivers. *Pure Applied Geophysics*, *179*, 1445–1460. https://doi.org/10.1007/s00024-022-02965-6

Nayak, S., Maity, S., Singh, K. S., Nayak, H. P., & Dutta, S. (2021). Influence of the changes in land-use and land cover on temperature over Northern and North Eastern India. *Land*, *10*(1), 52.

Nayak, S., & Tekemi, T. (2022). Assessing the impact of climate change on temperature and precipitation over India. In *Challenges and advance approaches for disaster risk reduction* (Vol. 42). https://doi.org/10.1002/joc.7332

Neog, P., Sarma, P. K., Borah, P., Sarma, D., Sarma, M. K., Borah, R., ... Chary, G. R. (2018). *Climate resilient experiences from NICRA implementation in North Bank Plains zones of Assam*. National Innovations on Climate Resilient Agriculture. All India Coordinated Research Project for Dryland Agriculture, Biswanath Chariali Centre, Biswanath College of Agricutlure, Assam Agricultural University.

Pachauri, R. K., & Meyer, L. A. (Eds.). (2014). *Climate change 2014: Synthesis Report. Contribution of working groups I, II and III to the fifth assessment report of the intergovernmental panel on climate change*. Intergovernmental Panel on Climate Change. www.ipcc.ch/report/ar5 Accessed March 12, 2016

Proctor, J., Haridasan, K., & Smith, G. W. (1998). How far north does lowland evergreen tropical rain forest go? *Global Ecology and Biogeography Letters*, 7(2), 141–146. https://doi.org/10.2307/2997817

Rahman, A. P. (2016, January). Climate change impacts Assam's cuisine and culture. *The Third Pole*. https://www.thethirdpole.net/en/climate/climate-change-assam

Rahman, M. A., Thomson, M. J., Shah-E-Alam, M., de Ocampo, M., Egdane, J., & Ismail, A. M. (2016). Exploring novel genetic sources of salinity tolerance in rice through molecular and physiological characterization. *Annals of Botany*, *117*(6), 1083–1097.

Ramesh, C. P., & Girish, B. S. (2022). Analysis of climate changes and slope instability factors triggered by natural rainfall. *International Journal of Scientific Research & Engineering Trends*, *8*, 374–378.

Rampini, C. (2016). Climate and energy justice along the Brahmaputra River in Northeast India. *Environment and Planning E: Nature and Space*, *5*(4), 1919–1946. https://doi.org/10.1177/25148486211032042

Rangnekar, D. V. (2006). *Livestock in the livelihoods of the underprivileged communities in India: A review*. International Livestock Research Institute (pp. 32, 45) http://www.ilri.org/Infoserv/webpub/fulldocs/LivestockInTheLivelihoods/LivestockUP_India_Final.pdf

Rao, M. G., Mukherjee, C. et al. (2011). Vision document for the state of Meghalaya 2030. *National Institute of Public Finance and Policy*. http://www.megplanning.gov.in/report/vision2030/ Accessed June 27, 2015.

Ravindranath, N. H., Rao, S., Sharma, N., Nair, M., Gopalakrishnan, R., Rao, A. S., ... Bala, G. (2011). Climate change vulnerability profiles for North East India. *Current Science*, (August), 384–394.

Roy, P. (2021). North East India saw largest share of the country's tree cover loss: Study. *Down to Earth*.

Sahoo, U. K., Singh, S. L., Sahoo, S. S., Nundanga, L., Nuntluanga, L., Devi, A. S., & Zothansiama, J. (2018). Forest dwellers' perception on climate change and their adaptive strategies to withstand impacts in Mizoram, North-East India. *Journal of Environmental Protection*, *9*, 1372–1392.

Saikia, P., Kumar, A., & Khan, M. L. (2016). Biodiversity status and climate change scenario in North East India. In *Environmental science and engineering* (pp. 107–120). Springer. https://doi.org/10.1007/978-3-319-31014-5_8

Saikia, Purabi, Kumar, A., Lal, P., & Khan, M. L. (2020). Ecosystem-based adaptation to climate change and disaster risk reduction in Eastern Himalayan Forests of Arunachal Pradesh, North East India. In *Nature-Based Solutions for Resilient Ecosystems and Societies* (pp. 391–408). Springer.

Salam, I. (2017). Human security—climate change—manipur: The gate way to South East Asia. *Journal of US-China Public Administration*, *14*(5). https://doi.org/10.17265/1548-6591/2017.05.005

Sangomla, A. (2021). Climate crisis in North East India: An outlier in Arunachal, Upper Siang now experiences increased rainfall. *Down to Earth*.

Shamsuddoha, M. & Choudhury, R. K. (2009). *Climate change induced forced migrants: In need of dignified recognition under a new protocol*. Campaign of Equity and Justice Working Group Bangladesh.

Sharma, D., & Saikia, S. (2021). Climate-resilient agriculture and food security in the light of climate-change with special reference to Arunachala Pradesh, North East India. *Zeichen Journal*, *7*(8), 166–188.

Sharma, Dinesh C. (2017a, December). Faced with climate change, Nagaland to revive traditional rice varieties. *Down to Earth*.

Sharma, Dinish C. (2017b). Why are we sweating in Shillong? *Down to Earth*.

Shimrah, T. (2017). Traditional ecological knowledge system as climate change adaptation strategies for mountain community of Tankhul Tribe in North East India. *International Journal of Environmental and Ecological Engineering*, *11*(7), 686–691.

Singh, K. S., & Kendra, K. V. (2021). Soil erosion control through bench terraces in Chandel District of Manipur. *J Krishi Vigyan*, *9*(2), 209–213.

Singh, M. K., Mahapatra, S., & Atreya, S. K. (2011). Solar passive features in vernacular architecture of north-east india. *Solar Energy*, *85*(9), 2011–2022. https://doi.org/10.1016/j.solener.2011.05.009

Stock, T. (1992). The convention on biological diversity 10 years on taking stock, looking forward. In *Convention on Biological Diversity Secretariat* (pp. 1–9). The Secretariat of the Convention on Biological Diversity

Stocker, T. F. et al. (2013). Technical summary. In *Climate change 2013: The physical science basis. contribution of working group I to the fifth assessment report of the intergovernmental panel on climate change* [T. F. Stocker, D. Qin, G.-K. Plattner, M. Tignor, S. K. Allen, J. Boschung, A. Nauels, Y. Xia, V. Bex, & P. M. Midgley (Eds.).]. (pp. 33–115). Cambridge University Press.

Subenthung Odyuo, N., & Nagesh, T. S. (2012). Fisheries and management status of Doyang Reservoir, Nagaland, North East India. *Indian Journal of Fisheries*, *59*(2), 1–6.

Taylor, M. (2018). Climate-smart agriculture: What is it good for? *Journal of Peasant Studies*, *45*(1), 89–107. https://doi.org/10.1080/03066150.2017.1312355

Thong, P., Thangjam, U., Sahoo, U. K., & Pebam, R. (2022). Socio-economic vulnerability assessment of shifting cultivators (Jhumias) amidst the changing climate in Mizoram, North East India. *Applied Geography*, *147*(October), 102790.

Thungon, P. (2020). Lukzepas A tribal societies of Arunachal Pradesh, North-east India. *Delhi Business Review*, *21*, 59–72.

TNN. (2018, September 5). Umiam water level reaches danger mark: Shillong news - Times of India. *The Times of India*. https://timesofindia.indiatimes.com/city/shillong/umiam-water-level-reaches-danger-mark/articleshow/65680744.cms

Tynsong, H., Dkhar, M., & Tiwari, B. K. (2020). Review: Traditional ecological knowledge of tribal communities of North East India. *Biodiversitas*, *21*(7), 3209–3224. https://doi.org/10.13057/biodiv/d210743

Upadhyay, A, Rai, R. K., Sarkar, S., Upadhyay, A. M., & Singh, V. P. (2009). GIUH based transfer functionfor Gomti River basin of India. *Journal of Spatial Hydrology*, 9, 24–50. *(PDF) Analysis of Rainfall Trends over Assam, North East India*. Available from: https://www.researchgate.net/publication/363554556_Analysis_of_Rainfall_Trends_over_Assam_North_East_India Accessed September 05, 2024.

Varah, F., & Varah, S. K. (2022). Indigenous knowledge and seasonal change: insights from the Tangkhul Naga in North East India. *GeoJournal*, *3*(11), 1–15. https://doi.org/10.1007/s10708-021-10559-3

Viswanathan, P. K., Kavya, K., & Bahinipati, C. S. (2020). Global patterns of climate-resilient agriculture: A review of studies and imperatives for empirical research in India. *Review of Development and Change*, *25*(2), 169–192. https://doi.org/10.1177/0972266120966211

Wisner, B. (2002). Wisner B. Who? What? Where? When? In an emergency: Notes on possible indicators of vulnerability and resilience: by phase of the disaster management cycle and social actor. *Environment and Human Security: Contributions to a Workshop in Bonn*, *12*(October), 7–12.

Yadava, A. K., Yadav, R. R., Misra, K. G., Singh, J., & Singh, D. (2015). Tree ring evidence of late summer warming in Sikkim, northeast India. *Quat Int*, *371*, 175–180.

Yengkokpam, S., Debnath, D., Bhattacharjya, B. K., Sarkar, U. K., Medhi, K., Gogoi, K., & Das, B. K. (2022). Exploring polyculture of small indigenous fishes with major carps in pens as a climate-resilient adaptation strategy for North Eastern wetlands of India. *Mitigation and Adaptation Strategies for Global Change*, *27*(January), 1–19. https://doi.org/10.1007/s11027-021-09981-2

Yepthomi, A. I. (2016, April 8). Projected impact of climate change in Nagaland. *Eastern Mirror*. https://easternmirrornagaland.com/projected-impact-of-climate-change-in-nagaland/

3 Status and Management of Water

Assam

Pallavi Sharma, Jyoti Prakash Deka and Runti Chaudhury

Water is one of the most important parameters to measure the environmental degradation of an area or an ecosystem. Water is the most indispensable requirement for all living organisms and any alterations in water may lead to the challenge of survival for these organisms, both human beings as well as animals. Good water quality is essential for all living organisms for their survival according to their needs such as drinking, bathing, habitat, and so on. The quality of water can be determined by studying the physical and chemical characteristics of the water. Because of vast population and the negligence of human beings the quality of water is deteriorating day by day. The detrimental effects caused by nature and human intervention has led to severe water challenges and water scarcity, which is posing a severe threat to society at large.

One of the major crises that society is facing today is "water scarcity." Water scarcity is described as a condition where water demand exceeds available water supply. A country or a region faces "water scarcity" when the availability of natural hygienic water falls below 1000 m^3 per person per year (Pereira et al., 2002; Dehghani et al., 2022). Water scarcity is not a concern with only a single living being at a particular location and time, it also affects a larger population within a certain geographic region (e.g., country) and pertains to larger timescales (years or months). It can occur because of natural low water availability as well as human-imposed activities to degrade the available natural water. The natural water body has the potential to renew or restore its originality. However, water pollution is the dominant factor to aggravate water scarcity by degrading the quality of water resources. Globally, 768 million people lack access to an improved water source, and more than 80% of these people live in rural areas (WHO and UNICEF, 2012).

Water scarcity arises in situations where there is insufficient water to simultaneously support both human and ecosystem water needs (White, 2014). Most often this arises because of a basic lack of water (i.e., physical water scarcity), but it may result from a lack of suitable water management infrastructure and policies. Recent estimates suggest approximately 4 billion people live under conditions of water scarcity for at least one month each year, with roughly 0.5 billion people exposed to severe water scarcity all year

DOI: 10.4324/9781003504870-4

round (Mekonnen and Hoekstra, 2016). According to estimates from the World Bank, the world could experience a deficit of nearly 40% between projected demand and available usable water as early as 2030 if population growth continues at the current rate and current water usage patterns and practices continue. Feeding 9 billion people by 2050 is predicted to result in a 60% increase in agricultural production, which may use 70% of the water resources available, leading to an increase in water withdrawals of up to 15%. The World Bank also states that, "About 40% of the world population live in water-scarce areas, and approximately ¼ of world's GDP is exposed to this challenge."'

By 2025, about 1.8 billion people will be living in regions or countries with absolute water scarcity. The increasing scarcity of water has resulted in efforts both internationally and domestically, particularly in developing nations, to advance a human rights–based approach for access to water (Narain, 2009). The United Nations Human Rights Council (UNHRC) declared that the right to water and sanitation already exists – as it is derived from the right to an adequate standard of living and is thus contained in existing human rights treaties – and that states had the primary responsibility to ensure full realization of this right (Gleick, 1998). The human right to water entitles everyone to sufficient, safe, acceptable, physically accessible, and affordable water for personal and domestic uses.

Audrey Azoulay, Director-General of UNESCO, says: "Let us state it clearly: this situation could bring about a systemic crisis in our societies. If humanity goes thirsty, fundamental questions on education, health and sustainable development will be sidelined, eclipsed by the daily struggle for water" (UN World Water Development Report, 2024).

India

India is no exception to water scarcity. The growing uncertainties of climate change have in fact made a bad situation worse; soon people may encounter unprecedented levels of water scarcity across the country. In India, the right to water has been derived from the fundamental right to life under Article 21 of the Constitution (Narain, 2009). The Composite Water Management Index (CWMI) of Niti Aayog has confirmed that 70% of India's water supply is contaminated. Globally, India is ranked 120th among 122 countries in Water Aid's water quality index (Desai, 2019). Seventeen percent of India's population does not have access to potable water, 80% of children suffer from waterborne diseases, and a total of 44 million people have illnesses related to poor water quality (Pant and Reddy, 2003). The human right to water and sanitation has evolved from soft to hard international law, where it is now considered as a "distinct, composite human right" (Meier et al., 2013). In 2010, the UN General Assembly recognized "the right to safe and clean drinking water and sanitation as a human right that is essential for the full enjoyment of life and all human rights" (UN, 2010).

North East India

Though earth is called the water planet as it is occupied by 75% of water, this water cannot be used for domestic purposes. Ocean water is saline in nature and is not fit for human consumption. Freshwater is just around 2.7% of the total water on the earth. North East India is blessed enough to have a bounty of accessible freshwater sources in the form of various rivers, streams, lakes, swamps, marshes, and so on, with the mighty Brahmaputra River along with its numerous tributaries bifurcating the whole area. These rivers are the lifelines of these regions acting like arteries in the human body and are supporting the social, ecological, cultural, and overall environmental setup. Additionally, these rivers along with their numerous wetlands, formed and fed by them, serve as the refuge to diverse organisms and sub-ecosystems (Bora and Goswami, 2017). During the last few decades, however, in the pursuit to catch up with rest of the world in terms of development, freshwater resources are continuously being tainted and deteriorated to an inconceivable stage. According to the Ministry of Environment, Forest and Climate Change (MOEFCC), 44 rivers in Assam have been affected by pollution. Some of them are Kopili, Digboi, Brahmaputra, Ronganodi, Deepor Beel, Bharalu, Digaru, Barak, Puthimari, Dihing, and others.

North East India has varied topography and climatic conditions which vary from tropical to temperate and alpine depending upon elevation. Flood plain areas have very different water regime, storage, and utilization practices compared to hilly regions.

The climate of North East India is changing as the rainfall patterns over the region have changed over the years, resulting in overall drying up. The climate change–induced water problem in the North East is due to the drying up of mountain springs in the region on which the states depend to meet their water needs. In Meghalaya, Mizoram and Manipur, 55.7, 54.6, and 54.4% of villages have springs. Nagaland and Arunachal Pradesh have 44.7 and 37.3% of villages with springs. The springs contribute a large share to the base flow of the large Himalayan rivers such as the Brahmaputra (Sangomla, 2021).

The physical factors responsible for water scarcity in the North East region include steep slopes due to the faults in the region causing overflow of rainwater and minimal water infiltration; impervious rocky surface preventing rainwater to infiltrate for groundwater storage; rainwater flowing down readily because of the great altitude and sloping nature of the region, assisted by gravitational force as well; and inaccessible groundwater because of insufficient stored groundwater due to the reasons cited.

Anthropogenic causes of water scarcity in the North East are decreased water supply due to rapid growth of population; increased affluence of population, which, according to Islam (2012), "has increased private car or bike ownership, creation of recreation sites like swimming pools and parks and their maintenance require huge amount of water". Other factors include expansion of industries, rapid urbanization, deforestation, global warming, and more.

The changing climate has caused significant uncertainty in urban water management and sustainability (Younos et al., 2019). The North Eastern region (NER) is characterized by a tropical monsoon climate with a rainy summer and dry winter, with annual rainfall 1857.2 mm, while in India it is 1083.2 mm (Sojitra and Pandya, 2015). The availability of net ground water in NER is 40.35048 billion cubic meters (BCM) (Kaur et al., 2022, Thapa et al. 2017). With the changing pattern of the monsoon rains, land degradation, and loss of forest cover, the water situation in the entire region is undergoing change, despite its bountiful natural environment.

Deterioration of Water

Water-related diseases, caused by unsafe drinking water and the absence of proper sanitation facilities, are among the leading causes of death in the developing world (Salman and Mclnerney-Lankford, 2004). Organic and biological impurities resulting from fertilizers, municipal wastewater, septic systems, and municipal and agricultural runoff are affecting the quality of the water of the Brahmaputra River daily. Around 97% of the domestic wastewater is untreated and discharged to the environment, with no proper arrangements of secondary and tertiary treatment facilities for treating and managing domestic wastewater. High levels of nutrients (nitrogen and phosphorus) cause eutrophication, leading to the growth of other microorganisms and causing contamination of surface water. Contaminants ingested into water supply causes many diseases. Examples of such pathogens are *Salmonella* spp., *Shigella* spp., *Vibrio cholerae*, and *E. coli* (Smruti and Sanjeeda, 2012, Choudhury et al., 2016). Faecal pollution of drinking water may introduce a variety of intestinal pathogens which may cause disease from mild gastro-enteritis to severe and sometimes fatal dysentery, diarrhoea, cholera, typhoid, hepatitis, giardiasis, and others. (Crown, 1986; Wanda et al., 2013; Choudhury et al., 2016). Because of certain highly toxic heavy metals in water like arsenic, some life-threatening diseases are reported such as cancer, hepatitis, and others. A few toxic substances often present in industrial waters, such as mercury and DDT, can undergo biomagnifications in the aquatic food chain.

Water contains different types of dissolved, suspended, and floating as well as microbial or micro bacterial impurities. The physical parameters for water are turbidity and colour, and chemical parameters for water are pH, electrical conductivity (E.C), total solids (TS), total dissolved solids (TDS), total suspended solids (TSS), total hardness, calcium hardness, and others. If the water has a high amount of any elements (e.g. temperature, TDS, and other chemicals) than the permissible limit, it may be harmful for all the aquatic living organisms that live in that water. Unsustainable water consumption practices by agricultural, industrial, and domestic purposes impose immense stress on water resources.

Water Demand

It is reported that more than 2 billion people are affected by water shortages in over 40 countries, 1.1 billion people do not have sufficient drinking water, and 2.4 billion people have no provision for sanitation. Current predictions forecast that by 2050, at least one in four people is likely to live in a country affected by chronic or recurring shortages of fresh water (UNESCO, 2003).

Water demand is defined as the volume of water requested by users to satisfy their needs. The consumption of water in a community is characterized by several types of demand. Water demands can be classified into domestic, industrial, institutional and commercial, public demand, and fire demand. Apart from this there is a percentage which goes as theft or loss. There is usually a wide variation in total water demand among different communities. This depends on population, geographic location, climate, the extent of local commercial and industrial activity, and the cost of water. Water use or demand is expressed numerically by average daily consumption per capita (per person). Per capita demand is dependent on factors like water supply system, cost of water, climatic conditions, size of the city, quality of water supply, pressure in the water distribution system, metering policy and charging method, industrial and commercial activities, and development of sewerage facilities.

There is great disparity in water demand of different countries. In the United States the average is approximately 380 litres (100 gallons) per capita per day for domestic and public needs. Overall, the average total demand is about 680 litres (180 gallons) per capita per day, when commercial and industrial water uses are included. However, in some developing countries it may average to as little as 15 litres (4 gallons) per capita per day. The world average is estimated to be approximately 60 litres (16 gallons) per person per day. The world population has more than tripled during the last century, while water uses for human purposes have multiplied sixfold (William & Rijsberman, 2000).

In Guwahati city, the average water consumption rate per day/per household is found to be 444.6 litres while the average water consumption per head is found to be 90.6 litres in (Bhattacharya and Borah, 2014). Thus, per capita water consumption is found to be less than the Indian urban standard of 135 litres (Vishwanath, 2013).

Water Management Strategies

Water management is defined as the process of planning, developing, and managing water resources in terms of both water quantity and quality across all water uses. It consists of the institutions, infrastructure, incentives, and information systems that aid and guide water management. A successful water management strategy ensures the availability of sufficient water of adequate quality to meet all water needs, including drinking, sanitation, agriculture and ancillary food production, energy generation, inland water transport, and so on, while also sustaining and protecting healthy water ecosystems and preventing devaluation of natural water bodies' aesthetic and spiritual values. The most important aspect of water management is water security for all. According to the World Bank, "water security is achieved when water's productive potential is leveraged, and its destructive potential is managed." Water security is a significant, escalating challenge for many countries, including India, and effective water management strategies are being increasingly touted as the means to achieve the end goal of water security. Proper planning and water resource management must focus on increasing capacity, adaptability, and resilience in the face of these challenges such as rapidly growing population and urbanization, climatic and non-climatic uncertainties, and so on.

The World Bank, the Government of India (GOI), and the various state governments have embarked upon several ambitious projects to streamline water management in rural India. The National Rural Drinking Water Program (NRDWP) guidelines of the GOI emphasize the involvement of the local government and communities in the planning, implementation, and management of drinking water supply schemes, along with capacity building and M&E systems. The plan aims to cover 90% of households with piped water and at least 80% of households with tap connections by 2022. About 75% of the Mahatma Gandhi National Rural Employment Guarantee Act funds have been reported to be utilized for water conservation efforts, complementing the Jal Shakti Abhiyan (JSA) launched in 2019 for rainwater harvesting, river rejuvenation leading to safe drinking water supply, and sanitation.

The Water Management program focuses on replenishing depleted underground aquifers and augmenting groundwater primarily with rainwater harvesting. This improves the availability and quality of groundwater in the long run and provides water security to rural households. Water resource augmentation structures include check dams, ponds, farm ponds, tanks, and recharge wells, among others. It emphasizes installation of rooftop rainwater harvesting systems in public buildings such as schools and institutional buildings that collect rainwater and store it in aboveground or underground tanks for later consumption. The water can also be recharged into the aquifers to create a local source of freshwater in the region. The Water Management program should also promote a variety of soil and water conservation measures to prevent soil erosion and salinity, and to maintain and improve soil fertility and productivity.

More than 80% of the hydropower potential of the North East region is located in Arunachal Pradesh (32 % of the national potential). The Dihang, Dibang, Kameng, Subansiri, Teesta, Kolong Kopili, and Lohit, along with upper Brahmaputra, are the main rivers identified for projects in the Brahmaputra basin. The aggregated irrigation requirements for the Brahmaputra basin are estimated to be 1.95 million hectare-meters for the kharif season and 3.25 million hectare-meters for the rabi season, while the total figure for the basin as a whole stands at 5.2 million hectare-meters (Goswami, 2004).

The Assam Scenario

Water use has increased rapidly over the past few decades, expanding at a rate that is more than twice as fast as the rate of population growth. The growth in human population and the diversification of human activities have led to an increase in the amount of water that is consumed. Water supply may be reduced because of rising per capita water demand in the home, agricultural, and industrial sectors. Additionally, it may significantly worsen the quality of the water.

There are abundant water resources in Assam. The breadth of its water supply is demonstrated by the size of its perennial rivers and other water bodies with rich aquifers. There are many different types of surface water, including rivers, streams, lakes, bogs, and ponds. Nearly the entire state has access to groundwater at shallow to medium depths. The Brahmaputra has very variable monsoon-driven hydrological regimes, with 60–80% of the total flow concentrated in 3–4 wet months. Although the availability of water resources varies seasonally and regionally, the annual availability is essentially constant. An estimated 5.44 billion cubic metres of groundwater are drawn annually, of which 4.85 billion are used for irrigation and 0.59 billion are used for home and industrial purposes. Groundwater is frequently extracted through drilled wells, tube wells, shallow tube wells, and deep tube wells in the state for irrigation, home consumption, and industrial use.

Surface water bodies occupy about 8251 sq km, or 10.5% of the state's overall geographic area. Approximately 6503 sq km of this region is covered by river systems, notably the massive Brahmaputra, and 1748 sq km by natural wetlands, including both man-made and seasonal waterlogged and marshy areas.

The Brahmaputra and Barak

The Brahmaputra River is aptly known as Assam's "Lifeline." One of the world's largest river systems, it influences the sociocultural landscape of Assam, other North Eastern states, and nearby nations like Bangladesh and China. It supports inland water transportation, irrigation, a richness of biodiversity, and Guwahati's main urban area, where it promotes growth and development. The Brahmaputra River is home to 20 wildlife sanctuaries and five national parks. Many endangered and threatened species, including mammals like the great one-horned rhinoceros, wild water buffalo, royal Bengal tiger, and Indian elephant, are found in this region's complex ecoregion, which is supported by numerous streams. The highly endangered Ganges river dolphin lives in the Brahmaputra River.

The Barak River System is the second largest in both Assam and North East India. The river drains a region of 52,000 sq km and is 900 km long from source to mouth. Originating in India, it travels 532 kilometres before reaching the border with Bangladesh. The Barak is a perennial river in the state, just like the Brahmaputra. Jiri, Siri, Madhura, Jatinga, and Larang are significant tributaries of the Barak River on its North bank. Sonai, Ghagra, Katakhal, Dhaleswari, Singla, and Longai are significant tributaries on the river's South bank. During the dry season, the Assamese rivers' flows significantly decrease. In the rainy summer months peak flow is maintained.

Assam has prospered thanks to the rivers' nourishment, and it has become home to plants and fauna that were once in danger of going extinct. The diverse cultural practices, means of subsistence, and daily routines of the populace have merged with those of the river. The rivers that run through Assam have had a very real impact on the lives of individuals who chose to live along its banks. The simple, perceptive, and peace-loving Assamese have known for ages how to navigate their adversary's undercurrents by creating low draught boats. By positioning themselves on the familiar end of the river and engaging the enemy strategically, the Assamese took advantage of the enemy, who arrived as Mughals and British invaders.

The river's generosity enables the people' entrepreneurial spirit to soar and the sand islands, or *chars* to grow. On *chars*, vegetables, mustard, paddy, lentils, corn, sugarcane, and groundnuts can be cultivated. These *chars* make excellent grazing areas for livestock because they quickly fill up with flora. Dairy is a large industry in Assam. The locals depend on the river for their daily needs and routinely catch fish. In this river basin, jute grows quickly, and the easy access to flowing water makes it easier to harvest the fibre and produce items made of jute. The silk business prospers in this humid environment along the banks because of the abundant mulberry trees used for silkworms.

Since ancient times, local communities have harvested river water in Assam. The cultural practices of Assam have water harvesting mechanisms. The Bodo community in rural Assam uses the "dong" method of traditional water management to direct river water into communities. Through earthen canals, water from the river is channelled towards the villages, where it is kept in local ponds. Such constructions are common along the Pagladiya River, which flows along Assam's Indo-Bhutan border.

Rainfall is the primary source of both surface and groundwater in the state. The South-West tropical monsoon, which rules from April to October with sporadic winter showers, has the greatest impact on Assam. The state experiences yearly rainfall ranging from 1780 to 3050 mm. Utilizing rainwater from the overwhelming amount of rain that Assam receives every year can be the answer to water crisis faced by people of Assam. The average rainfall in Assam every year is 1,722 mm per year which amounts to 5,64,816 million litres of water, enough to serve 37,65,440 people. However, residents of urban areas struggle each day to access their share of water. This largely owes to absence of rainwater harvesting in the state and the plan for the same. Unplanned city development leads more to floods than being able to use this rainwater. Rainwater harvesting has been a historically utilized method in India and has proved to be successful across several regions.

Wetlands

Wetlands rank among the most important ecosystems on Earth, because of the vast array of ecosystem services that wetlands provide. Wetlands play a significant role in maintaining many natural cycles, supporting a wide variety of biodiversity, and providing other services like regulating water flow, ground water recharging, and aiding in the purification and replenishment of water. For these reasons, wetlands are known as the "kidney of the ecosystem."

According to records, Assam has 690 lakes and ponds. These lakes and ponds have a surface area of 15,494.00 ha, or 0.20% of the state's overall land area and 15.30% of the total area covered by wetlands. The smallest one is 2.50 ha in size, while the largest one covers an area of 882.50 ha. Most of these wetlands have water that is not very murky. These lakes and ponds' aquatic vegetation has been examined, and the results show that most of them are either completely or sparsely vegetated. The Golaghat district has the highest number (113) of lakes and ponds, followed by the Dhubri (73) and Nagaon (68) (Status of Environment Related Issues: Assam ENVIS Centre, Ministry of Environment and Forests, Govt. of India n.d.). Deepor beel in Kamrup district, Dhir beel in Dhubri district, Tamaranga beel, and Dalani beel in Bongaigaon district are a few of the significant wetlands.

In addition to all the ecosystem services, it is a significant area for wetland agriculture, aquatic vegetation, and animals. These serve as a vital stopover for migratory birds and are home to a variety of waterbirds (Deka and Sharma, 2009). Because of the long food chain and abundant biodiversity that they support, wetland areas are frequently referred to as "ecological supermarkets" (Mitsch and Gosselink, 2000). As was previously mentioned, wetlands are crucial habitat for birds. Birds require wetlands for drinking water, feeding, resting, shelter, and other activities throughout their life cycle, especially during nesting.

Wetlands are among the most significant and productive ecosystems, yet they are among the most at risk. Thousands of acres of wetlands are being lost due to a variety of anthropogenic influences, such as climate change, population growth, urbanisation, and so on, instead of playing their essential functions in the ecosystem. Between 1970 and 2015, wetlands around the world lost almost 35% of their area, and since 2000, the pace of loss has been increasing annually (Simpson et al., 2021). The surviving wetlands are threatened by pollution, excessive use, invading species, and so on. According to Wetland International research coordinated by Wetlands International South Asia, India had the most wetlands in South Asia, accounting for around 4.6% of its total land area.

According to the Ministry of Environment and Forests, 42 of India's top 100 wetlands are threatened due to extensive encroachment and significant levels of human intervention (MoEF, 2007). In Assam, this critical aspect has not been given due importance and the wetlands are already in danger of losing their ecological character, mainly due to eutrophication. Incorporation of wetlands into an integrated river basin management plan is of utmost necessity for including those sites designated under the Convention on Wetlands of International Importance especially as waterfowl habitat (The Ramsar Convention Secretariat, 2014).

Pollution in Assam

In the North Eastern Region, numerous rivers pass by tea estates that make extensive use of pesticides and fertilizers. Water was revered by our forefathers because it is one of the essential elements for life. However, the price to be paid for this generation's lack of respect for water will be catastrophic. To preserve water and keep it safe, people's mindsets must immediately shift. Other causes of water scarcity include contamination, wasteful use, and improper use.

River contamination in Assam has recently become critically bad. The severe problem of Assam's river pollution was exposed by Prakash Javadekar, minister of the environment, forests, and climate change, in the Lok Sabha. Assam has the second-highest number of contaminated rivers after Maharashtra, according to statistics presented by Javadekar. The 44 Assam rivers that have become contaminated in recent years are Kundali, Kathakal, Baginadi, Dhanashiri, Barak, Palgladiya, Beki, Barabil, Bhogdoi, Desang, Digaru, Dihing, Haruchala, Kalioni, Puthimari, Ranga, Chamgudi, and Tengapukuri, among others, and were designated as "very polluted" rivers. The enormous population of humans and cattle that depend on river water makes this disturbing and hazardous. The primary cause of this form of pollution is the untreated waste that is discharged into rivers by industry.

River contamination is made worse by wetlands degradation. A network of inflow and outflow channels connect Deepor Beel (a Ramsar wetland reserve and animal sanctuary on the outskirts of Guwahati) to the Brahmaputra. Numerous endangered species, including the greater adjutant stork, red-headed vulture, and white-rumped vulture, are found in the swamp. Deer and wild elephants are frequently spotted in the reserve. The area has long been utilized as Guwahati's waste dump, putting its rich biodiversity in danger. The Bharalu River carries untreated sewage into it. The water quality at Deepor Beel was Priority IV in certain spots and Priority V in others, according to assessments made by the Pollution Control Board (PCB) in 2017 (I represents the cleanest, and V represents the most contaminated, on a scale of I to V). Deepor Beel is on its way to extinction.

Sewage Mechanism in Assam

The largest city in Assam, Guwahati, through which the Brahmaputra flows, generates about 154 million litres of sewage per day. The increased sewage generation is due to the city's population growth, which has expanded at a

20% average yearly rate. The primary cause of river pollution is the untreated direct discharge of sewage and wastewater into the Brahmaputra. There isn't a sewage treatment plant in Assam, according to a nationwide inventory of sewage treatment plants (STPs) published in March 2021 by India's pollution authority, the Central Pollution Control Board. Assam is the most populous state without STPs and smaller states and island territories (Nagaland, Manipur, Meghalaya, Andamans, and Lakshadweep) also lack them in India.

The Brahmaputra is becoming more polluted by sewage. Not all polluted locations are under the watchful eye of the Assam Pollution Board. Unregulated small industrial and manufacturing facilities discharge several sewage kinds. Domestic and municipal trash is channelled into the river through several drains. All these sources' untreated sewage eventually finds its way to the Brahmaputra.

The Cultural and Spiritual Significance of Water

Water's cultural significance stems from its deep link to spirituality in numerous cultures and traditions. "Cultural" water is regarded as sacred since it is linked to legends of creation and spiritual beliefs. It might come from a variety of sources, including rivers, lakes, springs, or wells, and its value is frequently linked to its location or history. Rivers are significant in many cultures, including the Nile River in ancient Egypt and the Ganges River in India. These rivers are frequently considered sacred because they are seen as the land's lifeblood, providing water for agriculture and sustaining local communities.

Water is fundamental to many religious and spiritual beliefs around the world. It is regarded as a source of life, symbolizing rebirth. Water cleans and purifies the body, and these two attributes give water a highly symbolic role. Water is consequently an important ingredient in ceremonial and religious rites. Water is frequently seen as a God, Goddess, or divine agency in religions. Rivers, rain, ponds, lakes, glaciers, hailstorms, and snow are some of the forms that water can take when interpreted and incorporated into cultural and religious context. Religious water is thought to have the power and capacity to reform the world, eliminate sin, and create holiness.

River worship is a phenomenon seen in cultures all across the world, notably among the indigenous tribes in North East India. It is observed among adherents of all major religions. All major Hindu groups regard these rivers as sacred, just as the Brahmaputra in Assam. Though river worship is not uncommon across civilizations, the worship of the Kopili river in Assam is significant since it is carried out by three distinct indigenous groups of people with different traditions and faiths, speaking multiple languages. In the hill districts, the three major indigenous communities of Jaintia, Karbi, and Dimasa inhabit the bank of the Kopili.

There are, however, formal differences in the rites performed by these three tribes. All the three of these communities worship the Kopili. According to Jaintia history, their forefathers crossed this river on their route to their current home in the hills. The river is revered as the mother Goddess, and sacrificial rites are done every year to appease her. They refuse to cross the river unless they conduct sacrifices to the Goddess, and they are certain that any disobedience will result in a curse from the Goddess. Similarly, the Dimasa and Karbi tribes revere the river. Propitiation is carried out for well-being of an individual or a family. Such propitiation takes place at two levels – voluntary and curative. In the former any individual or head of a family may personally carry out propitiation rituals for personal or familial well-being in the form of good harvest, general health, job prospects, and so on. On the other hand, divination is involved in the curative form, such as divination for curing illness in a member of a family, caused by the wrath of the river deity for some ill action on behalf of the victim.

In Assam rituals not only use water as a divine form but certain phrases are spoken in description of the process, such as *pani tula* for bringing water ceremonially during auspicious occasions, *pani kota* during divination, and so on.

Water is the lifeline of cultures, civilizations, various floral and faunal species, and what is commonly referred to as the natural ecosystem. Water provides nutrients in the form of silts and nourishes the valley's vegetation, whether tamed or wild. Human communities have used their intimate understanding of local ecosystems to tap river resources and establish social institutions. The Ahom hydraulic management system, as well as the traditional "dong" irrigation system used by the Dimasa and the Bodo, are just a few examples from this region. Such modest interventions are vastly different from modern megastructures, which disrupt the normal flow of the river and inundate broad areas, destroying biomes with unfathomable ecological consequences.

"Karbi people have been maintaining a harmonious relationship with water since ages. They have long held this belief that water is essential for their overall well-being and for environment sustainability," said Maggie Katharpi, a faculty at Diphu Government College. She further said, "... Embedded in their traditions and beliefs, water lore encompasses the spiritual, cultural, and ecological aspects of Karbi people's connection with this life-giving element" (Times of India, January 18, 2024).

Looking Ahead

People's Participation

Human attitude can reflect how individuals perceive their immediate environment (Swim et al., 2009). A fixed mindset would view the environment as static and unchanging, and as a result, it would have a detrimental impact on a person's propensity to engage in pro-environmental behaviour, whereas adopting pro-environmental behaviours is more likely when one has a mindset towards growth, which sees the environment as changeable and capable of improvement (Soliman and Wilson, 2017a).

When people see themselves as a part of the natural world that they may potentially pollute, they tend to have attitudes of environmental care (Schultz and Zeleny, 2000). Such attitudes may be established through awareness and information; otherwise, a lack of comprehension of the importance of taking sustainable and environmental action leads people to engage in degrading behaviours (Imran et al., 2014).

Some people are obstinate when it comes to problems with water management and desire a judicial system to punish offenders. Regular awareness efforts can aid in the development of a social norm, which would then mould people's attitudes and behaviour in favour of the management of the water resources.

Collaboration

Collaboration among different agencies, like local government and line agencies working together with academic institutions and educational institutions to share information, develop curricula and encourage students to participate in socially responsible and environmentally friendly activities would pay dividends in the long run. Several initiatives may be put into place for community engagement in water management. Community involvement can result in shared attitudes, knowledge, behaviours, and norms surrounding effective water management. It can assist in establishing a norm of social responsibility. Large amounts of rainwater can be used for other purposes. For new structures and flats, a policy must make rainwater harvesting mandatory. There should be increased awareness of the crisis' magnitude. Like this, water used for residential reasons can be frequently recycled and used for rituals, customs, and basic tasks in rural areas like bathing or washing animals and clothes in rivers.

Legal Measures

A larger need exists for stricter implementation of legal measures (in the form of fines for those who break the law), which can aid in promoting a change in mindset in favour of personal accountability for achieving better water quality.

People treat the environment in whatever way they choose when they think they are in control of it and have a right to it. So long as there are weak laws and no established, regularly evaluated law enforcement, this gets worse. Later, these presumptions get ingrained in people's thoughts, influencing their attitudes and behaviours, and eventually becoming the social norm (Schwartz, 1977). Since human activities are the primary cause of problems with water quality, managing water quality will largely depend on understanding the potential obstacles to and enablers of changing human attitude and behaviour.

It's the collective duty of all to protect and cherish the only home that we have ever known, the Earth, and in doing so all need to act in a systematic, scientific, and unified way so that the future generations can survive and prosper in a world where drinking water is accessible to everyone.

Conclusion

India is home to approximately 17% of the world's population but only 4% of the world's fresh water. Fresh water plays a pivotal role in every regional issue like food security, energy production, economic development, and poverty reduction. It can reflect cultural traditions and interactions as seen with the mighty Brahmaputra in the state of Assam and how belief systems and knowledge have been rooted in history and traditional perspective. It reflects the fabric of social life. But it also remains a commodity that lacks the required attention. Society seems to be blindfolded when it comes to water issues, however, the apparent neglect cannot be justified by lack of knowledge.

Water and society are deeply interwoven, as humans have contributed to bringing major changes to the water cycle with the increasing impact of climate change and global warming. The change in the rainfall pattern has started affecting agricultural productivity thereby impacting the economy. In due course, population growth, shifting economic conditions, unsustainable social activities, and injudicious technological advancements will create greater and more multifaceted stress on water resources. Availability and its continuous exploitation are leading to shortage of this scarce resource thereby having a cascading impact on social and economic activities.

However, as water is considered a human right, it is important that all human beings have equitable access to water. It exists both as a physical object and a non-physical entity, and as an instrument in the life of every human. The mere existence of water therefore brings into question dominant theories of property and management, theories fundamental to most discussions about society, but taken for granted in current mainstream research.

It is of utmost importance that improved water management strategies are adopted for adaptation, mitigation, and resilience in a rapidly changing world. It is important that effort is made to understand this physical aspect of water and track how societies adapt to it. Sustainable management practices such as recycling of water, rainwater harvesting, water resource budgeting, groundwater recharge, micro irrigation, wetland protection, water body restoration, conservation agriculture, and other nature-based solutions that can help conserve and protect water resources must be formulated and implemented. Participatory measures where all sections of the society benefit are needed to be adopted where people's intimacy with water is encouraged at the cultural level, so that their emotional bond with water is sustained, along with scientific measures.

If society, industry, and ecosystem desire to get all the water they need for different purposes, it's time to re-establish humans' relationship with water. The fight against the issues which are affecting water resources must be fought jointly as only by acting together it can be sustained for the future. The involvement of the authorities and all sections of society are required, as well as both finance and planning, which should work hand in hand for achieving the sustainable development goals.

References

Bhattacharya, P., & Borah, R. (2014). Drinking water in Guwahati city: Its past, present status and associated problems. *Space and Culture, India*, *1*(3), 65–78.

Bora, M., & Goswami, D. C. (2017). Channel morphology and hydraulic geometry of River Kolong, Nagaon district, Assam, India: A study from the standpoint of river restoration. *Current Science*, *113*, 743–751.

Chaudhury, S., Ockenhouse, C. F., Regules, J. A., Dutta, S., Wallqvist, A., Jongert, E., Waters, N. C., Lemiale, F., & Bergmann-Leitner, E. (2016). The biological function of antibodies induced by the RTS; S/AS01 malaria vaccine candidate is determined by their fine specificity. *Malaria Journal*, *15*(1), 1–12.

Crown, G. F. (1986). *Water-borne diseases in the United States*. CRC Press, Inc.

Dehghani, S., Bavani, A. M., Roozbahani, A., Gohari, A., & Berndtsson, R. (2022). Towards an integrated system modeling of water scarcity with projected changes in climate and socioeconomic conditions. *Sustainable Production and Consumption*, *33*, 543–556.

Deka, D. R., & Sharma, P. (2009, August). Values of wetlands: A critical analysis on wetland environment degradation in the Digaru and Kalong River basins of Assam, India. In N. Deka (Ed.), *Proceedings of UGC sponsored National Seminar on 'Strategy for Resource Functioning of India's North East: A Geographical Perspective'* (pp. 56–61). NN Saikia College.

Desai, D. (2019). Quality of water in India's state capitals. *orfonline.org*. https://www.orfonline.org/expert-speak/quality-of-water-in-indias-state-capitals-58308

Gleick, P. H. (1998). The human right to water. *Water Policy*, *1*(5), 487–503.

Goswami, D. C. (2004). Flood forecasting in the Brahmaputra River, India: A case study. Available at www.southasianfloods.org

Imran, S., Alam, K., & Beaumont, N. (2014). Environmental orientations and environmental behaviour: Perceptions of protected area tourism stakeholders. *Tourism Management*, *40*, 290–299.

Kaur, H., Choudhary, S., Kumar, N., & Jat, M. K. (2022). Conservation of traditional water bodies; A study of Gopal Sagar and Roop Sagar, Deeg, Rajasthan.

Meier, B. M., Kayser, G., Amjad, U., & Bartram, J. (2013). Implementing an evolving human right through water and sanitation policy. *Water Policy*, *15*(1), 116–133.

Mekonnen, M. M., & Hoekstra, A. Y. (2016). Four billion people facing severe water scarcity. *Science Advances*, *2*(2), e1500323.

Ministry of Environment and Forests. (2007). Conservation of wetlands in India: A profile (approach and guidelines). MoEF.

Mitsch, W. J., & Gosselink, J. G. (2000). The value of wetlands: Importance of scale and landscape setting. *Ecological Economics*, *35*(1), 25–33.

Narain, V. (2009). Water as a fundamental right: A perspective from India. *Vermont Law Review*, *34*, 917.

Pant, H. K., & Reddy, K. R. (2003). Potential internal loading of phosphorus in a wetland constructed in agricultural land. *Water Research*, *37*(5), 965–972.

Pereira, L. S., Oweis, T., & Zairi, A. (2002). Irrigation management under water scarcity. *Agricultural Water Management*, *57*(3), 175–206.

Salman, S., & Mclnerney-Lankford, S. (2004). *The human right to water: Legal and policy dimensions*. World Bank.

Sangomla, A. (2021). Climate crisis in North East India: What is behind water scarcity in the region. https://www.downtoearth.org.in/news/climate-change/climate-crisis-in-NorthEast-india-what-is-behind-water-scarcity-in-the-region-78910

Schultz, P. W., & Zeleny, L. C. (2000). Promoting environmentalism. *The Journal of Social Issues*, *56*, 443–457.

Schwartz, S. H. (1977). Normative influences on altruism. In *Advances in Experimental Social Psychology* (Vol. 10, pp. 221–279). Academic Press.

Simpson, M., Stroud, D., McInnes, R., Galewski, T., Segura-Champagnon, L., Dinesen, L., ... & Rebelo, L. (2021). Global wetland outlook: Special edition 2021. Secretariat of the Convention on Wetland.

Smruti, S., & Sanjeeda, I. (2012). Microbiological analysis of surface water in Indore, India. *Research Journal of Recent Sciences*, *1*, 323–325.

Sojitra, M. A., & Pandya, P. A. (2015). Comparative study of daily rainfall forecasting models using adaptive-neuro fuzzy inference system (ANFIS). *Current World Environment*, *10*(2), 529–536. https://doi.org/10.12944/cwe.10.2.19

Soliman, M., & Wilson, A. E. (2017a). Seeing change and being change in the world: The relationship between lay theories about the world and environmental intentions. *Journal of Environmental Psychology*, *50*, 104–111.

Status of Environment Related Issues: Assam ENVIS Centre, Ministry of Environment and Forests, Govt. of India (n.d.).

Swim, J., Clayton, S., Doherty, T., Gifford, R., Howard, G., Reser, J., ... & Weber, E. (2009). *Psychology and global climate change: Addressing a multi-faceted phenomenon and set of challenges. A report by the American Psychological Association's task force on the interface between psychology and global climate change*. American Psychological Association.

Thapa, R., Gupta, S., Reddy, D. V., & Kaur, H. (2017). An evaluation of irrigation water suitability in the Dwarka river basin through the use of GIS-based modelling. *Environmental Earth Sciences*, *76*(14), 1–12.

The Ramsar Convention Secretariat. (2014). Ramsar.

UN World Water Development Report 2024. UN. (2024). https://www.unwater.org/publications/un-world-water-development-report-2024

UNESCO. (2003). Water for people, water for life: The United Nations world water development report. (World Water Assessment Program), 10 (UNESCO Publishing 2003).

United Nations (UN). (2010). Human rights obligations related to access to safe drinking water and sanitation. Note by the secretary-general. A/65/254. United Nations.

Vishwanath, S. (2013, February 15). How much water does an urban citizen need? *The Hindu*, available at: http://www.thehindu.com/features/homes-and-gardens/how-much-water-does-an-urban-citizen-need/article4393634.ece (accessed 12 December 2013)

Wanda, E. M., Gulula, L. C., & Phiri, A. (2013). Hydrochemical assessment of groundwater used for irrigation in Rumphi and Karonga districts, Northern Malawi. *Physics and Chemistry of the Earth, Parts A/B/C*, *66*, 51–59.

White, C. (2014). 28. Understanding water scarcity: Definitions and measurements. *Global Water: Issues and Insights*, *161*.

WHO & UNICEF (2012). Progress on Drinking Water and Sanitation: 2012(Update), USA. http://www.unicef.org/media/files/JMPreport2012.pdf [Accessed: 20.02. 2013].

William, J. C., & Rijsberman, F. R. (2000). *World water vision—Making water everybody's business* (World Water Council), 4. Earthscan Publications Ltd.

Younos, T., Lee, J., & Parece, T. (2019). Twenty-first century urban water management: the imperative for holistic and cross-disciplinary approach. *Journal of Environmental Studies and Sciences*, *9*(1), 90–95.

4 Climate Change and Tribal Livelihood of North East India

Somenath Bhattacharjee

As long as the human population was small, during the hunter-gatherer and the early settled land-use phase of Homo sapiens, they remained integrated well within the boundaries of the ecosystem or landscape. With the process of settlements appearing with the advent of agriculture about 10,000 years ago, the exploitative phase of the human population started growing rapidly. After the industrial revolution, though a levelling-off of population has occurred in the industrialised world, it continues to grow rapidly in the developing world. Closely linked to this phenomenon is the great economic divide that has happened and is continuing to grow between the developing and developed world. All this has resulted in a situation where a small section of the global population of a few developed nations has an exploitative hold over the global natural resource base. In the process this minority is marginalising a substantial section of the human population living in the developing world. It is in this context that one must consider climate change and the whole set of issues closely linked to it, broadly referred to as global change.

What Is Global Change?

The term global change is often misunderstood and interchanged with climate change, as if the two were synonymous (Ramakrishnan, 2001). There is much more to global change than mere climate change related issues. Climate change can be viewed from two different perspectives: that which is a natural phenomenon, caused through geological events, as has happened in the earlier geological history of our planet. This is beyond human control. Climate change in the present context is only concerned with human-induced climate changes brought about through industrial emissions into the atmosphere and the consequent changes in the atmospheric elemental composition.

Global change is not merely related to the climatic change or geomorphological changes alone. Related to climatic changes, the significant changes which are taking place in the livelihood and daily practices of human life, particularly in those societies who are deeply attached to the forest and natural resources, cannot be overlooked or isolated. Although the concept of global change has arisen out of research on climate change it has a more

DOI: 10.4324/9781003504870-5

holistic perspective and includes all aspects of human activities and life affected by climate change.

Biological invasion, the colonisation by exotic species in an alien environment, is a phenomenon which has played havoc in the past and continues to do so; water hyacinth in water bodies or Lantana invasion on the land are examples of such an invasion in India (Ramakrishnan, 1991). With the introduction of technologies for rapid and mass transport of humans and materials, the problem of biological invasion on a global scale is becoming more and more serious. This is difficult to control once it has occurred. Apart from the economics of it, there are ecological and social problems that stand in the way of control of biological invasion. Such factors often have a direct impact on the floral and faunal resources of any local ecosystem, which are directly related to the food, rituals and traditional health care practices of the concerned societies dwelling in that ecological habitat.

Linked to all this is biodiversity depletion, another serious global change issue (May, 1992) that is totally irreversible. Biodiversity once lost cannot be recovered at all, as seen from the accelerated large-scale human-induced extinction of species.

Climate Change

Climate change, as noted earlier, is perhaps the best publicised component of global change. For the enhanced 'greenhouse effect' due to trapping of long-wave radiation emitted from the earth's surface, industrial activities of humans are implicated. Though global warming as a phenomenon is well documented and human activities are linked to this change, its precise impact at a local/regional scale is still uncertain, due to the difficulties involved in teasing out natural climatic variability from the observed trends. There is also some uncertainty about relating climate change with human influences, as climate is subject to natural variability, and due to the difficulties in separating human-induced impacts from the background noise of climate fluctuations (Walker et al., 1999). However, based on a large body of available evidence and predictive modelling analyses, the Inter-governmental Panel on Climate Change (IPCC, 1996) concludes that there exists discernible human influence on global climate.

With a predicted temperature change in the range of 2 to 3°C, Hulme and Viner (1998) suggest the following changes in the Indian sub-continental region: an increase in rainfall, in some areas up to 50%; a possible reduction in the dry season length by several months; and decreased rain in other areas. Local variability is even more difficult to predict. The uncertainties on the impact of climate change will get further complicated through interaction between other global change related elements. It is in this context that the probable impacts of climate change on ecosystems must be viewed.

Very few studies are available on ecosystem impacts in the Indian context. Much less is available on the interactive impacts of climate change when viewed in the context of other global change impacts. Even less is available in the context of the human dimensions of the problem, since the culture to combine biophysical with the socio-economic dimensions of the problem is much less researched. The author had the advantage of looking at global change from a holistic perspective, linking the issues involved in the context of the sustainable livelihood/development concerns, when he got involved in a socio-ecological analysis of the socio-cultural landscape of the North Eastern hill regions of India, starting from the early 1970s (Ramakrishnan, 1992 and over 200 research publications). Since then, the author has extended the studies to other Himalayan and Western Ghat mountain regions (Ramakrishnan et al., 2000), based on the experiences gained from the North Eastern hill region.

Impact of Climate Change

Mountain societies are directly dependent upon forests and the resources contained therein through extraction of several non-timber forest products and indirectly through the linkages with their traditional agricultural systems. The economy of these traditional societies, being driven by biodiversity, and both natural and human-managed climate change–related impacts taken along with other anthropogenic impacts on ecosystems will threaten their immediate livelihood concerns, with long-term impacts on sustainable development of the region (Ramakrishnan, 2001).

Land use and cover changes driven by human activities are by far the most dominant component of global change (Ramakrishnan, 1998). A variety of human activities like timber extraction, conversion of forested land for agriculture, reclaiming wetlands for urban development or other purposes, destruction of mangroves for marine aquaculture along the coastal zone, and so on, could be implicated as factors responsible for land use and cover change.

Placed in the context of climate change, the combined impacts with anthropogenic factors make the net outcome more complex. At best only calculated conjectures can be made, as the author has tried to do for the North Eastern situation. The impact could be felt in a variety of different ways: weed invasion particularly by exotic species, biodiversity depletion, soil erosion, loss of soil carbon and nutrients, increased release of radiatively active trace gases to the atmosphere, all collectively influencing local/regional climate and ecology. Important consequences of all this are the general slowing down in ecosystem recovery processes and enormous cost in terms of money and labour to be incurred in the rehabilitation of degraded landscapes.

The understanding of the land use–land cover relationships, from the point of global change issues, remains inadequate. To improve this understanding, land use and cover change must be linked to human actions. These actions are the product of individual/group behaviour within a given socio-economic and environmental setting (Ramakrishnan, 1998). Land use change-based human adaptation, in turn, leads to environmental feedback and additional human responses. The ability to arrive at generalisations at this biophysical-social interphase is still poorly developed.

An excellent example of the interphase feedback is that derived from the shifting agricultural landscape in the North Eastern hill region of India (Ramakrishnan, 1992). Proximal driving forces such as population pressure, biodiversity, soil fertility depletion and otherts have many underlying key drivers: policy driven socio-economic setting, determining the market forces and the linked institutional mechanisms interacting in a variety of different ways to determine the land-use dynamics. These pressures decline as one moves away from the urban centres in North East Indian uplands. It is to be noted that the *Jhum* system is replaced by rotational fallow system, leading to sedentary farming when one moves closer to the urban centre (Adapted from Ramakrishnan, 1992).

In the context of the shifting agricultural landscape as a whole, not only in North Eastern India but also elsewhere, the tendency of the policy planners and development agencies has always been to implicate the shifting agriculturist as the cause of deforestation and land degradation. What is seen is that, in the ultimate analysis, large-scale timber extraction due to external forces arising out of larger policy decisions of the governmental agencies is the chief cause of land degradation. At a local level, the interphase feedback exhibits themselves as a series of responses by the people to altered ecosystem conditions both in time and space. These adaptations are based on adaptive evolution in their traditional ecological knowledge (TEK)–based land-use practices, linked to both natural and agricultural systems. In a regional context, such land-cover changes influence bio-geo-chemical processes, the feedback from which may affect the land-use dynamics itself and the social setting in which they operate.

It is crucial to understand the complex human adaptations to cope with the impacts of biological invasion, for designing strategies for its management, if not total control. Ultimately the overarching objective should be to consider sustainable livelihood concerns of local communities in the short run and sustainable development of the region as part of the long-term strategy, based on an adaptive landscape management plan (Ramakrishnan, 2001).

Coping with Environmental Uncertainties in the Context of Climate Change

The contemporary situation of global warming, continuous environmental degradation and climate change has a certain impact on the livelihood, economic and cultural sustainability of different ethnic groups, who are deeply attached to their local ecological niche. It is the age-old indigenous knowledge and traditional practices that used to protect and preserve their local ecological floral and faunal resources, which ultimately sustains their livelihood and culture. This can be elaborated through the following discussion.

Shifting Agriculture

Large areas of forest on all three tropical continents have a variety of forms of shifting agriculture, also known as 'swidden', 'slash-and-burn' and 'rotational bush-fallow' agriculture and by a variety of local names for centuries. In India, it is commonly referred to as *Jhum*. The basic principle of this extensive form of agriculture is the alternation of short crop phases (usually one or two years of cropping) with phases of natural (or slightly modified) vegetation fallow. Yield is thus managed on a long-term basis, rather than by maximisation in the short-term.

Shifting agriculture systems traditionally maintain diversity in the cropping phase by utilising mixed cropping, the perennial shrubs and trees being separated in time and confined to the fallow regenerative phase of the forest, in a temporally separated agro-forestry system. Agricultural ecosystem functions such as nutrient cycling and pest population dynamics are controlled both through the complex cropping and the fallow phases. The key to the stability of the system thus lies in retaining a minimum agricultural cycle length (length of the fallow period before the farmer returns to the same site or another cropping phase). It is during the fallow phase that soil fertility recovery occurs, which in turn determines economic yield.

In the North East Indian hill areas, where shifting agriculture is the major land use, a minimum of 10-year cycle was found to be necessary for its sustainability – both economic and ecological efficiency – as evaluated through energy efficiency and fertility sustainability (Ramakrishnan, 1992). However, the average cycle length is less than 5 years at present.

Shifting agriculture, though complex in terms of biodiversity, generally tends to be managed casually. The number of crop species in the mixture may vary considerably, from six to over 40, depending upon the agricultural cycle. Linked to traditional animal husbandry such as poultry and swine, and with the forest sector, this land use is highly complex, with differences in details, such as agricultural procedures, cropping and yield patterns. Consequently, the ecological and economic efficiencies of these systems may differ considerably (Ramakrishnan, 1992). Crop diversity alone does not express the total biodiversity in the system. The species diversity in a plot may go up to about 60 or more if weeds are considered. When the fallow phase is included, the number may run to more than a hundred.

In the recent past, because of increasing population pressure, there has been a shift from more extensive to more intensive systems of land use in many parts of the tropical world. In North Eastern India where the shifting agricultural cycle has come down below 5 years, as has happened amongst village communities around city centres, long fallow systems have been replaced by shorter fallow systems. The herbaceous weed biomass of the short fallow phase may be slashed and allowed to decompose in the surface layers of the soil before the land is prepared for cultivation. The cropping intensity may vary from place to place depending upon the situation. Very often, three to ten species of leguminous crops may form part of the mixed cropping system involving traditional crops such as rice or maize along with lesser-known crops of food value. *Flemingiavestita* in North East India is a lesser-known legume crop species, used for inter-cropping as well as for crop rotation once in 3–4 years; it helps in improving soil fertility through nitrogen fixation.

Tropical Compound Farming

One of the oldest traditional ways in which humans in the humid tropics imitated nature in their agricultural practices is through incorporation of trees and other perennials as components of an elaborately constructed home garden, a system still prevalent in many areas of the humid tropics. Home gardens are small plots, usually of one half to two hectares, located close to the habitation and fertilised with household wastes. They are rich in plant species diversity (30 to 100 species), dominated by woody perennials and structurally stratified, with a mixture of annuals and perennials of varied habit herbs, shrubs, trees and vines as well as more conventional food crops. The farmer obtains food products, firewood, medicinal plants, spices and ornamentals, and some cash income all the year round.

Among tropical compound farming systems, the more specialised fields often include 'high value' crops, which form part of traditional economies, being bartered for other products. As more market structures have developed, these crops have become important components of the cash economy. Many, although not all, of these initial cash crops are perennial, including many fruit trees, bananas, and others, as well as such crops as ginger, pineapples, yams and special produce like the cultivation of broom grass (for broom making) or bamboo (for a variety of purposes) (Ramakrishnan, 1992).

Valley Agriculture

Valley agriculture is practised throughout the hill terrain, both at low and high elevations of the North Eastern region of India, which is a sedentary form of wet rice (*Oryzasativa*) cultivation and is a complementary system to *Jhum*. It is done wherever the terrain permits, on flat lands between hill slopes. However, small terraces may also be built by some tribal communities all around the flat valley land extending into the foot of the hill slopes. This results in a saucer-shaped structure in which rice plots are organised. Obviously, this cropping system is restricted due to topography.

The soil in valley lands is fertile due to nutrient washout from the hill slopes, and therefore does not need added fertilisers. The main advantage is that the land gives sustained yield year after year, unlike the *Jhum* system that is under cropping only once in a few years of fallow interval depending upon the cycle. The *Apatani* version of wet rice cultivation is one of the most advanced. With elaborate spatial organisation of crop varieties so as to differentially exploit the nutrient and water gradient that exists in their system which is partly based on village waste recycling, this wet rice cultivation that combines pisciculture with it is one of the most advanced, with an exceptionally high energy and economic efficiency (Ramakrishnan, 1992). This offers opportunities for redevelopment, with additional scientific inputs.

Contemporary Agro-Ecosystem

In recent times, several studies show there is a large body of evidence accumulating to suggest that biodiversity within agro-ecosystems plays an increasing role in determining agro-ecosystem stability and resilience. A common perception about agro-ecosystems is that their composition is deficient in biodiversity and their structure and function impaired under human management. This is a limited view based on perceptions arising from agricultural management of the northern hemisphere. Such a perception of agriculture is becoming more and more obvious in the developing world too, under the impact of the 'green revolution'.

The rich diversity in agro-ecosystem types that existed not too long ago is being lost rapidly, and along with it, the crop diversity at the specific and sub-specific levels. Biodiversity contributes in a variety of ways to ecosystem functioning, such as production, decomposition, nutrient cycling dynamics, and thus, towards stability and resilience of the system. It is generally acknowledged that ecosystem complexity decreases as habitats change from forest to traditional agriculture and modern agriculture. If total biodiversity (not just that which is associated with the production system itself) is plotted at each point along this gradient, it is highly probable that the resulting relationship will be monotonic and decreasing (Swift et al., 1996), the exact form of the curve being somewhat uncertain.

At very low intensities of management the emphasis is on short-term increased production levels through external energy subsides – at the expense of biodiversity, ecosystem stability and resilience. This suggests that biodiversity is little affected by initial stages of management and that the loss is gradual until some critical stage in management intensity is reached, followed by a rapid decline, perhaps somewhere between low and middle intensities of management. The North East Indian case study illustrates this (Ramakrishnan, 1992). It seems to follow that planning research activities should be focussed on this critical gradient in management.

While a variety of models for loss in biodiversity under varied intensities of management regimes for agriculture are proposed, it seems to follow from recent studies that biodiversity decline is sharp somewhere in the area close to the middle intensity of management. If that be so, it is crucial to have a level of management that is closer to this critical area for sustaining biodiversity in agriculture at the same time to optimise production.

Possible Pathways for Agricultural Development

If we consider high-input modern agriculture, considered as an auto-route, as only one of the possible pathways for agricultural development, one could have at least two more additional pathways for sustainable agriculture: (a) evolution by incremental change or (b) restoration through the contour pathway. The first is a natural process in relation to the bio-cultural process of human evolution. It is the process where simple technologies are still used to pursue primitive agriculture as a process of traditional economic organisation and as an oral tradition. Still, this is a type of surplus economy, practised by different ethnic groups of India. The second pathway is directly related to complex technology-based modern agriculture. One important aspect of this agriculture in the contemporary period is organic cultivation with an eco-friendly and environmentally sustainable pathway and procedure to conduct the entire process.

The 'auto-route' pathway is well known because this is the direction which much of the research efforts of the agricultural scientific community have so far taken. An appropriate metaphor would be the engineer who plans an auto-route by drawing a straight line from one point to another on a map and proceeds to build a straight and level road regardless of physical impediments. Such an agro-ecosystem type would stand apart as an artificial entity from the rest of the landscape as an attempt to convert the natural ecosystem into one that contains only those biological and chemical elements that the planner desires, almost irrespective of the background ecological conditions.

There are many strategies for the sustainable pathways for improving the system of land use and resource management in North Eastern India. The following strategies suggested by Ramakrishnan et al. are based on a multidisciplinary analysis and many have already been put into practice:

- Wide variations in cropping and yield patterns under *Jhum* practised by over a hundred tribes under diverse ecological situations, where transfer of technology from one tribe/area to another alone could improve the *Jhum*, valley land and home garden ecosystems. Thus, for example, emphasis on potato at higher elevations compared to rice at lower elevations has led to a manifold increase in economic yields despite low fertility of the more acid soils at higher elevations.
- Maintain a *Jhum* cycle of a minimum of 10 years (this cycle length was found critical for sustainability when *Jhum* was evaluated using money, energy, soil fertility biomass productivity, biodiversity and water quality as currencies) through greater emphasis on other land-use systems such as the traditional valley cultivation or home gardens.
- Where *Jhum* cycle length cannot be increased beyond the 5-year period that is prevalent in the region, redesign and strengthen the agro-forestry system by incorporating ecological insights on tree architecture (e.g. the canopy form of tree should be compatible with crop species at ground

level to permit sufficient light penetration and provide fast recycling of nutrients through fast leaf turnover rates). Local perceptions are extremely important in tree selection, for introduction into the cropping and fallow phases of *Jhum*, as is being done in a major initiative in the state of Nagaland in North East India.

- Improve the nitrogen economy of *Jhum* in the cropping and fallow phases by introducing nitrogen-fixing legumes and non-legumes. A species such as the Nepalese alder (*Alnusnepalensis*) is readily taken in because it is based on the principle of adaptation of traditional knowledge to meet modern needs. Another such example is the less-known food crop legume *Flemingiavestita*, traditionally used by the tribals as an important species when *Jhum* cycles decline below 5 years. Some of the important bamboo species, highly valued by the tribals, can concentrate and conserve important nutrient elements such as N, P and K. They could also be used as windbreaks to check wind-blow loss of ash and nutrient losses in water.
- Speed up fallow regeneration after *Jhum* by introducing fast-growing native shrubs and trees.
- Condense the timespan of forest succession and accelerate restoration of degraded lands based on an understanding of tree growth strategies and architecture by adjusting the species mix in time and space.
- Redevelop village ecosystems through the introduction of appropriate technology to relieve drudgery and improve energy efficiency (cooking stoves, agricultural implements, biogas generation, small hydroelectric projects, etc.). Promote crafts such as blacksmithery and products based on leather, bamboo and other woods.
- Strengthen conservation measures based on the traditional knowledge and value system with which the tribal communities can identify, for example, revival of the sacred grove concept based on cultural traditions which enabled each village to have a protected forest once upon a time, although few are left now.

These considerations formed the basis for decentralised village development in the North Eastern hill state of Nagaland in India. Over a thousand villages in the state of Nagaland have been organised into Village Development Boards (VDBs). The highly distorted shifting agriculture which indeed is basically an agro-forestry system, but is now operating at or below subsistence level, is being redeveloped by strengthening the tree component that has been weakened due to extreme deforestation in the region. The entire basis for this incremental build-up is the rich traditional ecological knowledge base of these hill societies. The project implementation by the Nagaland government officials through VDBs aims at augmenting the traditional system of agriculture rather than changing it radically. The Nepalese alder-based agroforestry systems, with tree planting done both in space and time (during the cropping and fallow phases of shifting agriculture) and maintained for hundreds of years by some of the local tribes like the Angamis, gave the impetus to this initiative.

Traditional tribal societies in North Eastern India organise nutrient-use–efficient crop species on the top of the slope and less efficient species along the bottom to match with the soil fertility gradient on a steep slope. With shortening of the shifting agricultural cycle the farmer tends to emphasise more on tuber and vegetable crops as compared to their emphasis on cereals under longer cycles. Operating under a mixed cropping system, where the species are sown at the same time soon after the first rain during the monsoon, the farmer harvests crops sequentially as and when the crop matures over a period of a few months; after harvesting the economically useful component, he recycles the biomass into his agricultural plot, which decomposes rapidly.

The whole of the Nagaland experiment is based upon effective forest fallow management through assisted secondary succession. Fallow management as an issue of renewed interest arises out of three major considerations:

- Fallow management emphasises the close linkage that always exists between agriculture and forestry at the landscape level of the TEK evolved by the mountain societies.
- The whole issue of finding sustainable livelihood for local communities made up of marginalised farmers in the developing tropics, as part of a short-term strategy for sustainability of agriculture with improved yields. The shifting agricultural situation in India is one such typical example where this land-use system which has fallen to disrepute has to be pulled up for improved and sustainable agriculture.
- Fallow management system of agriculture could be viewed as part of a landscape management strategy for coping with issues linked to 'global change' land use and cover change, depletion of biodiversity, land degradation, desertification and of course global warming and climate change.

Realising that biodiversity and ecosystem complexity do contribute in a variety of ways to ecosystem functions and that agro-ecosystems do harbour a great deal of biodiversity valuable for general human welfare, it is reasonable to go in for a mosaic of natural ecosystems co-existing with a wide variety of agro-ecosystem models derived through all the three pathways. In a landscape mosaic, agro-ecosystems form one of the components within a wide variety of ecosystems. Natural systems such as forest, grassland and freshwater pond or lake are interspersed with human-made agricultural monocropping systems such as rice, wheat or maize fields, village woodlots and so on. The relative area apportioned for each of these land-use units would of course be determined by ecological and social location specificities. Such a highly diversified landscape unit is likely to have a wide range of ecological niches conducive to enhancing biodiversity and at the same time ensuring sustainability of the managed landscape itself. The more recently evolved biosphere reserve concept of UNESCO, indeed a rediscovery of the concept of the sacred landscape of traditional societies (Ramakrishnan, 2001) dating back to antiquity, is an attempt towards such an integrated management strategy to conserve natural resources for sustainable use with inter-generational equity concerns.

Traditional Practices in the Context of Climate Change

It is in the context of climate change one must study the traditional sustainable living practices to learn from them and update them. With this in view a few examples from the North East are given and a few problems have been mentioned that have arisen because of changes without preparing the people to cope with them. The first are land management and pesticide practices. Another is the traditional water management system.

Spraying dry ash on brinjal leaves in the cultivation season is a practice in Assam. This is an insecticide meant to protect the plant from insects and pests. Chemicals and pesticides were unknown to the traditional farmers. They found their own pesticides just as they selected the best fruit and preserved it for seeds for the next season. One does not fail to notice the hill terracing practices among the Angami and Chakhesang tribes in, respectively, the Kohima and Phek districts of Nagaland. The terraces have been preserved for more than a thousand years. This contrasts with the slash and burn (*Jhum*) practices among many other hill tribes. These are two practices carried out side by side. There are strong opinions against *Jhum* based on environmental and ecological concerns.

During a village-level consultation on sustainable agriculture and organic manure in Chokihola in the Nilip Block of Karbi Anglong, the role of dry leaves was discussed. An interesting reply from many participants was that leaves are a waste material and need to be destroyed by burning. When they were convinced that leaves are an important source of soil nutrition, they took a resolution not to burn them but to use them for organic manure to enrich the soil.

It is to be mentioned here that, in the case of *Jhum* cultivation as noticed among the Karbi of Karbi Anglong, Assam, the concerned people use the portion of land totally based on their indigenous knowledge. In the *Jhum* cultivation the hilly tract is selected for the cultivation, and they cultivate for one or two years in that place and then change the land of cultivation the next year. So, it is regarded as the shifting hill cultivation, because they change the cultivable land every year. They do so because after a year or two the land will lose its natural fertility and production would decrease. Generally, the *Jhum* cultivation is performed for one year at a place. With the emergence of new economic organisation, the people concentrate on increasing production, resulting in *Jhum* cultivation at a place for more than a year, due to the decreasing of natural forests. The field is left to grow into a jungle for a number of years. When the jungle grows for numbers of years then the land gradually gets back its natural fertility. Presently very little land is left for *Jhum* cultivation. Therefore, people leave the land to grow into a jungle for only two or three years, then again cultivate that land every three or four years.

The phases of alternation of land during shifting cultivation can be represented through Figure 4.1.

A portion of land or resource area, like 'A' is cultivated for a period of 2 to 3 consecutive years. Thereafter it is shifted to some other nearby or distant plot of land like 'B', and such shifting usually means going to higher altitudes of land in general. In the same way plot 'C' and 'D' will be selected and shifted after a period of 2 to 3 consecutive years of farming. Gradually over the period the concerned people will go back and select that particular plot of land which was selected by them at the earliest, that is, 8 to 12 years ago.

When the cultivation work was done in the lower altitudes (Plots D and C) due to the availability of larger cultivable land the quantity of production was high, and it was more variable in nature. The concerned people used to fulfil their daily livelihood requirements through the barter exchange method. However, in the higher altitudes (Plots B and A) the amount of production usually becomes less due to comparatively smaller area of cultivable land.

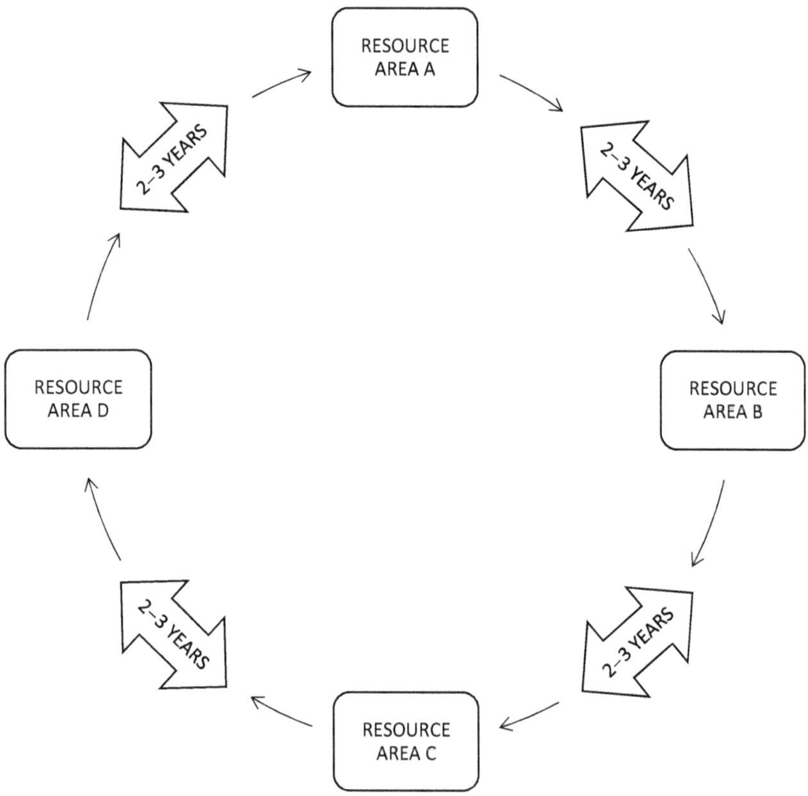

Figure 4.1 Indigenous framework of shifting cultivation or *Jhum* cultivation.

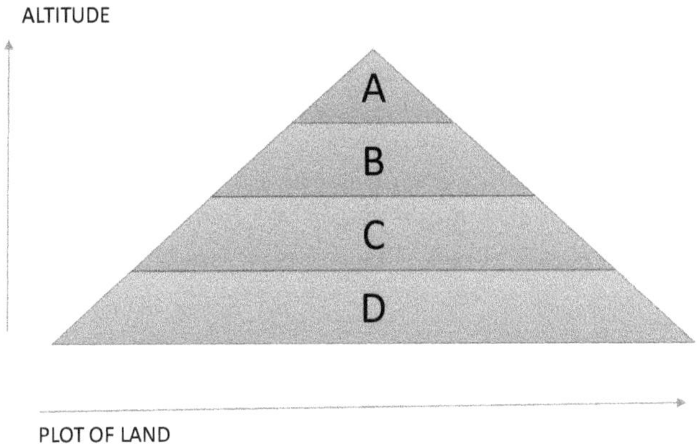

Figure 4.2 Spatial selection of land with the change of altitude.

The variability of production is lower too. In such cases the Barter exchange method was practised with the cultivators settled in the plains. Among the Karbi the method of Barter exchange and distribution of productions were collectively done with community consensus to maintain the egalitarianism in society. Although the village headman was present during the process of distribution he had no claim of some special privilege on the production.

A noticeable consequence of *Jhum* cultivation is the change in settlement with changing of *Jhum* land. The *ang-uiarit* (the dense forest area where the big trees are located) has more naturally fertile land and so people always want to select *ang-uiarit*. The *ang-uiarit* gives more production. In those areas the herbs and shrubs grow much less. People in search of *ang-uiarit* as cultivable land got further away from their settlement with passing years. As the *Jhum* fields got further away from their settlement the problem of communication arose. It was very difficult to perform the work by travelling far distances and it was difficult to protect their crops from wild animals. After harvesting it becomes difficult to carry their crops from the *Jhum* field to the household. Due to these numerous problems some people changed their settlement with the change in their *Jhum* lands.

Moreover, measures need to be taken on these issues and to mitigate the impact of floods in Assam. Floods have become an annual feature here and the monsoons of 2000 were among the worst. During a Disaster Preparedness and Mitigation Training programme, many farmers of Golaghat District suggested that, to cope with the floods, the traditional Assamese mud floor houses need to be changed to houses on stilts as in use among the Mishing. They live in the flood plains and have adapted their houses to this need. The farmers added that one may also need to change the schedule of cultivation of the main food grains. That may make the facility of irrigation for cultivation available to many more but may also require cultural changes among the people to facilitate a new cropping pattern and schedule.

In Search of Solutions

Environmental concerns are growing with demographic changes, modernisation, and introduction of the market economy. Different actors are in search of logical and practical solutions to these problems. Also, the reorganisation of the North East based on linguistic and ethnic groups has had a direct impact on the lives of the communities in this region. Some of the solutions suggested by the government and by private parties are:

Official Solutions: The finance minister of India declared a policy on the National Watershed Development movement during the last two budget sessions. It is worth looking at the implications of the movement on the lives of common people. The policy speaks about starting a Watershed Development Fund with National Bank for Agriculture and Rural Development (NABARD) to cover 100 priority districts within 3 years. The central government will provide necessary matching assistance to NABARD. The implementation of the scheme will be through the community infrastructure for micro watershed projects. They include the gram panchayats and self-help groups. On the surface the scheme looks good. But the hard fact remains that the government will no longer provide irrigation facilities. Finances from NABARD to gram panchayats and other bodies for micro watershed projects imply that the users, that is, farmers, will have to pay back to NABARD, probably with interest, in cash. It implies that villages opting to join the proposed movement will not only look after the operational and maintenance cost but also be taxed for the hard work they put in. A one-time management subsidy and recurring assistance will be provided to 'all registered Water User Associations' (WUA) that are linked to incremental water rate collection. The state governments that will substantially increase the rates of water and the WUAs taking responsibility for the collection of 'incremental' water rates will be bestowed with favours. By implication, those who do not fall in line shall suffer. This policy implies that in the proposed National Movement of Watershed Development, irrigation will be regarded as a 'service or utility' and the cost of the same along with interest must be shifted from the State to the individual user.

People's Solutions: On the other side, many communities have tried their own solutions with partial or full success. These exercises have given hope and encouragement to others to take up watershed management and other programmes in the vulnerable areas of the North East. Seeing some of these successful demonstrations, many others in the region are taking up similar programmes. They need to be studied to find the direction of the changes that must be introduced as a solution to ecological degradation that is basic to climate change. A few examples are given in the following.

An example of a search by the people concerns a stream named Pouchaiki between Asalu and Impoi villages in the Mahur subdivision of North Cachar Hills, Assam. During the author's visit to Impoi in 1995, the local village management committee voiced their concern about the decreasing flow in the stream which is their main source of water. After discussion the village authorities decided to declare the entire catchment of the stream a protected area. They took a resolution accordingly and prohibited cutting of trees and killing of birds and animals in the catchment area. Fruit-bearing trees like orange were planted. Six small check dams were built through the food-for-work programme. Even though the village did not implement the resolution in its entirety, the action had its impact. A recent review of the stream reveals that its flow has increased especially during the dry period from February to April. Now the people have sufficient water for their needs.

In Jongpha village of Karbi Anglong, the Rengma Nagas have a similar successful watershed experience. During the project preparations for the International Fund for Agriculture Development/North Eastern Council, the satellite imagery maps had identified in Assam, the Mahur area of NC hills and Hamren in Karbi Anglong, Ukhrul district of Manipur and Dadenggre and West Garo Hills in Meghalaya as the most environmentally degraded areas. Perhaps the solutions to the problem can be thought of by involving the people whose lives are at stake and directly at the receiving end of wrong practices.

Conclusion

That there is a problem of climate change is clear today. But to find a solution to the problem one must understand the context in which the people live. For sheer survival the poor tend to intensify the problem created by the rich. A solution to it cannot be found by those in power taking a decision to change the system. The examples given in this chapter show that unless value-loaded pro-people, pro-poor positioning is taken the results will be unsatisfactory. It means that capacity building of the facilitating actor, who may be a voluntary worker or a government servant, must be according to the needs of the people (or the context) and not vice versa. Otherwise, one becomes inward looking, focusing on the structure. Involving the people (context) in the process of development is extremely important and vital. People's ownership over the process will ensure success of any intervention targeted to improve the quality of holistic life.

Bibliography

Agarwal, A., & Narain, S. (1991). Technology control, global warming and environmental colonialism: The WRI report. *Social Action*, *41*(1), 1–28.

Bhattacharjee, J. B. (1990). Social and political institutions of the Angami Nagas. In J. Sarkar & B. D. Ray (Eds.), *Social and political institutions of the hill people of North East India* (pp. 102–116). Anthropological Survey of India.

Census of India. 1991. Nagaland State District Profile 1991. Census Publications.

Das, N. K. (1987). *Ethnic identity, ethnicity and social stratification in North East India*. Inter-India Publications.

Deeney, J., & Fernandes, W. (1992). Tribals: Their dependence on forests, their traditions and management systems. In W. Fernandes (Ed.), *National development and tribal deprivation* (pp. 49–75). Indian Social Institute.

Elwin, V. (1969). *The Nagas in the Nineteenth Century*. Oxford University Press.

Epao, V. (n.d.). *From Naga Animism to Christianity*. Published by the Author.

Fernandes, W., Menon, G., & Viegas, P. (1988). *Forests, environment and tribal economy: Deforestation, impoverishment and marginalisation in Orissa*. Indian Social Institute.

Fernandes, W., D'Souza, N. G., Choudhury, A. R., & Asif, M. (2001). *Development-induced displacement, deprival and rehabilitation in Andhra Pradesh 1951–1995: A quantitative and qualitative study of its extent and nature*. Indian Social Institute and Guwahati: North Eastern Social Research Centre.

Government of Assam. (1993). *Assam Forests at a Glance*. Assam Forest Department.

Haimendorf, F. V. (1992). *Tribes of India: The struggle for survival*. Oxford University Press.

Hulme, M., & Viner, D. A. (1998). Climate change scenario for the tropics. *Climate Change*, *39*, 145–176.

Hutton, J. H. (1969). *The Angami Nagas*. Oxford University Press. First published in 1921 by Macmillan & Co.

IPCC. (1996). *Climate change 1995: Contribution of working group I to the second assessment report of the intergovernmental panel on climate change*. J. J. Houghton, F. L. G. Meiro, B. A. Calendaer, N. Harris, A. Kattenber, & K. Maskell (Eds.). Cambridge University Press.

Iralu, E. (2000). *Folk elements in Achebe: A comparative study of Ibo culture and Tenymia culture*. Ura Academy.

Laya. (1999). *Land alienation in Tribal Andhra Pradesh*. Laya Office.

Marwa, I. S., & Srivatsava, V. K. (1992). Khel gate social structure: A study of their relationship and a note on the place of material culture in Anthropology. In S. M. Channa (Ed.), *Nagaland: A Contemporary Ethnography* (pp. 75–120). Cosmo Publication.

May, R. M. (1992). Past efforts and future prospects towards understanding how many species there are. In O. T. Solbrig, H. M. van Emden, & P. G. van Oordt (Eds.), *Biodiversity and global change* (Monograph 8, pp. 71–81). International Union of Biological Sciences.

Mehrotra, N. (1992). Angami Naga women: Some reflections on their status. In S. M. Channa (Ed.), op. cit. pp. 147–180.

Mooney, H. A., Cushman, J. H., Medina, E., Sala, O. E., & Schulze. E. D. (1996). *Functional roles of biodiversity: A global perspective*. John Wiley, SCOPE Volume 55.

NABARD. (n.d.). https://www.nabard.org/demo/auth/writereaddata/File/21_Watershed_Management.pdf

Pathy, J. (1988). *Ethnic minorities in the process of development*. Rawat Publications.

Peseyie, K., & Imchen, T. (1984). *Nagaland: Paniketi on mountain slopes*. Department of Agriculture, Government of Nagaland.

Ramakrishnan, P. S. (1984). The science behind rotational bush-fallow agriculture system (*Jhum*). *Proceedings of Indian Academy of Science (Plant Sciences)*, *93*, 379–400.

Ramakrishnan, P. S. (1991). *Ecology of biological invasion in the tropics*. National Institute of Ecology and International Science Publications.

Ramakrishnan, P. S. (1992). *Shifting agriculture and sustainable development: An interdisciplinary study from North Eastern India*. Parthenon Publications and Paris.

Ramakrishnan, P. S. (1998). LUCC initiatives in the Asian region: Lessons from Indian Experience. *LUCC Newsletter*, *3*, 9–14.

Ramakrishnan, P. S. (2000). Ecology teaching in India and in developing countries. *Biology International*, *39*, 33–44.
Ramakrishnan, P. S. (2001). *Ecology and development*. Book Trust of India (in press).
Ramakrishnan, P. S., Purohit, A. N., Saxena, K. G., & Rao, K. S. (1994a). *Himalayan environment and sustainable development*. Indian Natural Science Academy, Diamond Jubilee Publications.
Ramakrishnan, P. S., Campbell, J., Demierre, L., Gyi, A., Malhotra, K. C., Mehndiratta, S., Rai, S. N., & Sashidharan, E. M. (1994b). *Ecosystem rehabilitation of the rural landscape in South Central Asia: An analysis of issues*. Special Publication, UNESCO (ROSTCA).
Ramakrishnan, P. S., Das, A. K., & Saxena, K. G. (1996). *Conserving biodiversity for sustainable development*. Indian National Science Academy.
Ramakrishnan, P. S., Saxena, K. G., & Chandrashekara, U. M. (1998). *Conserving the sacred: For biodiversity management*. UNESCO, Oxford and IBH Publications.
Ramakrishnan, P. S., Chandrashekara, U. M., Elourd, C., Guilmoto, C. Z., Maikhuri, R. K., Rao, K. S., Sankar, S., & Saxena, K. G. (2000). *Mountain biodiversity, land use dynamics and traditional ecological knowledge*. UNESCO, Oxford and IBH Publications.
Rao, Subba, et al. (Eds.). (n.d.) *Andhra Pradesh Paryavarana Paristhiti: Praja Nivedika 1990*. Academy of Gandhian Studies.
Raven, P. H. (1985). Disappearing species: A global tragedy. *The Futurist*, *19*, 8–14.
Ravindranath, N. H., & Sukumar, R. (1998). Climate change and tropical forests in India. *Climate Change*, *39*, 563–581.
Reddy, N. S. (1988). Depriving tribals of land: Andhra move to amend land transfer laws. *Economic and Political Weekly*, *23*(29), 1458–1461.
Sala, O. E., Chapin III, F. S., Gardner, R. H., Lauenroth, W. K., Mooney, H. A., & Ramakrishnan, P. S. (1999). Global change, biodiversity and ecological complexity. In B. Walker, W. Steffen, J. Canadell, & J. Ingram (Eds.), *The terrestrial bio-sphere and global change* (pp. 304–328). Cambridge University Press.
Sanyu, V. (1996). *A history of Nagas and Nagaland*. Commonwealth Publishers.
Swift, M. J., Vandermeer, J., Ramakrishnan, P. S., Anderson, J. M., Ong, C. K., & Hawkins, B. (1996). Biodiversity and agroecosystem function. In H. A. Mooney, J. H. Cushman, E. Medina, O. E. Sala, & E. D. Schulze (Eds.), *Functional roles of biodiversity. A global perspective* (pp. 261–298). John & Wiley, SCOPE Series.
UNEP. (1992). *The state of the environment 1972–1992: Saving our planet*. United Nations Environment Programme.
Vyasulu, V. (1998). The South Asian model. *Paper Presented at the International Conference Colonialism to Globalization: Five Centuries after Vasco da Gama*. Indian Social Institute.
Walker, B., Steffen, W., Canadell, J., & Ingram, J. (1999). *The Terres-trial biosphere and global change: Implications for natural and managed ecosystems*. Cambridge University Press, Synthesis Volume, IGBP Book Series 4.

5 Environmental Attitudes and Behaviour among Mizo Adolescents

Laldinpuia, Rinpari Ralte and C. Zothanmawia

A sustainable future is one in which resources are not destroyed through either overconsumption or pollution, and in which it is ensured that a high quality of life is available for future generations (Oskamp, 2000). To secure such a future the present-day youth need to have beliefs, values, and attitudes which are pro-environmental. While there has been a shift in how individuals view environmental resources, it is seen that the movement for conservation of the environment has been spearheaded by young people such as Greta Thunberg, who has taken the issue to the world stage. There are others, such as Xiuhtezcatl Martinez, Nyombi Morris, Licypriya Kangujam (from Manipur, North East India), Xiye Bastida, Lesein Mutunkei, and others, who together are leading the way on global climate action.

Exploring how the young people of Mizoram, a state in North East India, view the environment and their opinions, emotions, and behaviour towards the environment is, would make a suitable starting point to figuring out where society is headed in terms of a sustainable future.

Mizoram

Mizoram, the 23rd state of the country, is a North Eastern state bordered by the states of Assam to the north, Tripura to the north west, and Manipur to the north east, as well as the countries of Bangladesh to the west and Myanmar to the east and south (Lodrick, 2023). Mizoram is well-known for being one of the greenest states of India. In terms of forest coverage per geographical area, Mizoram has the highest forest coverage in India with 84.53% of its landmass providing a natural habitat for a diverse range of species of flora and fauna (Forest Survey of India, 2022).

The Mizo forefathers believed themselves to be spawns of nature, which is evident in their worship of '*khua*' (or nature) and their belief in fate being determined by nature as a force ('*khuarel,*' which translates to 'determined by nature'). They believed the cosmos consisted of the corporeal world and the intangible world of spirits, with humans and animated lifeforms possessing spirits (Vanlalruata, 2017). For the typical Mizo, nature is not just the surroundings nor just a resource, but something deeply connected to the

DOI: 10.4324/9781003504870-6

Mizo way of life. The ancestors of the Mizo people believed in a spiritual force residing within nature, shaping their worldview and influencing their daily lives (Bhattacharjee, 2017; Hmingthanzuali & Pande, 2009; Lalthangliana, 2005; Shakespear, 1912; Zairema, 1989). Although the Mizo people have abandoned their traditional religion, their past beliefs are still evident in contemporary society through various linguistic expressions and concepts; for example, terms such as '*khuanu malsawmna*' ('Mother Nature's blessings'), '*vanneihna*' ('good luck', where '*van*' is associated with heaven), and '*vanduaina*' ('bad luck') and many others show that the Mizo people believe that their future and their own being is determined by a force outside of their own control. Moreover, the deep connection that the Mizo people have with nature goes beyond mere spiritual beliefs, as nature evokes a strong emotional resonance within them. The concept of '*khawtlang lunglen*' is a standard example of this sentiment (Zuali, 2012), describing a feeling of longing or melancholy for the past and association with the hills and mountains that define their landscape. This emotional yearning is strongly linked to their geographical and cultural roots and reveals the deep attachment the people have to the land that has shaped their identity over generations.

Modernization has prompted the Mizo people to gradually detach themselves from their profound connection with nature. Urbanization and infrastructure development have led to migration from rural to urban areas, diminishing direct interaction with the natural environment (Nag, 1993; Hmingthanzuali & Pande, 2009; see also Saitluanga, 2018). Technological advancements and changing economic activities have shifted reliance away from traditional practices deeply rooted in nature, while Westernization and globalization have influenced lifestyle choices, often prioritizing materialism over environmental stewardship (Chew & Sarabia, 2016). Environmental degradation resulting from rapid industrialization further weakens the tangible link between the Mizo people and their surroundings, exacerbated by an educational system that prioritizes academic subjects over traditional ecological knowledge (Lalruatpuii & Ngurtinkhuma, 2023). This distancing from nature poses significant challenges for preserving Mizo cultural heritage and maintaining sustainable relationships with the environment. Yet, the natural environment across the state (including the state capital, Aizawl) is such that the people are surrounded by a verdant landscape which is an integral component of their existence in that environment, thereby perhaps having an impact on their perceptions, beliefs, emotions, and attitudes towards the natural environment.

Mizoram is a land of hills, characterized by a temperate climate due to its elevated terrain, which results in moderate temperatures throughout the year (JICA, 2015). The website of the Government of Mizoram, Environment, Forests & Climate Change Department (https://forest.mizoram.gov.in/page/forest-types) identified various forest types within the state, primarily categorized by altitude, rainfall patterns, and dominant species composition. These include tropical wet evergreen forest, Montane subtropical forest, temperate forests, bamboo forests, Quercus forests, and Jhumland. The Mizoram government has established ten protected areas covering approximately 8% of the state's total geographical area. These include the Dampa Tiger Reserve, Murlen National Park, Phawngpui National Park, and several wildlife sanctuaries. These protected areas are home to a remarkable variety of flora and fauna, and the forests and lands serve as habitats to numerous mammal species, including the state animal, the red serow, as well as tigers, leopards, clouded leopards, Asiatic black bears, and sun bears, among others (Department of Environment, Forests, & Climate Change, 2020a).

Despite significant conservation efforts, Mizoram faces contemporary environmental challenges due to deforestation, pollution, and resource scarcity. Felling and mining pressure, forest fires, encroachment, developmental activities, and floods are all exhibiting an upward trend in Aizawl over the past 30 years (Department of Environment, Forests, & Climate Change, 2020b). However, while the prevalence of Jhum cultivation has decreased, the environmental damage caused by it persists, with its lingering impact resulting in patchy deforestation, soil erosion, soil nutrient depletion, reduced soil moisture retention, and a decline in native biodiversity. The changed Jhum cultivation pattern, with reduced cycle of cultivation and fallow periods, has contributed to the drying up of springs and rivers, along with the depletion of underground water reserves (Leblhuber & Vanlalhruaia, 2012; Rawat et al., 2017, as cited in Department of Environment, Forests, & Climate Change, 2020b). Deforestation contributes to an increased risk of flash floods, although large-scale flooding hasn't been reported yet.

These ecological degradations have significant consequences on forest cover and overall environmental health. According to the 2022 data from Global Forest Watch, Mizoram has experienced substantial losses in primary forest and tree cover over the past two decades. Besides deforestation and habitat loss, other anthropogenic activities such as the increasing number of vehicles, poorly planned urbanization, population growth, and inadequate sewage systems have led to further increase in environmental issues (Renthlei, 2016). These activities have led to problems like air and noise pollution, disrupting the region's ecological balance and impacting human health. Moreover, the combined effects of local factors and climate change have intensified environmental challenges in Mizoram. In such a scenario, delving into environmental behaviour and the psychological factors behind it would perhaps unfurl the processes why people behave in the ways that they do towards the environment.

Environmental Behaviour

The field of environmental psychology has witnessed a surge of interest in recent years regarding the relationship between environmental attitudes and behaviours. Environmental attitude refers to how individuals perceive and evaluate the environment, including their beliefs, emotions, and values regarding nature, sustainability, and ecological matters. Such an attitude often shapes people's understanding of the environment and influences their intentions and actions. According to Azjen's theory of planned behaviour (Ajzen, 1991), attitudes act as significant determinants of behavioural intentions, which in turn guide actual behaviours. Consequently, individuals harbouring positive environmental attitudes are more likely to embrace and enact pro-environmental behaviours (Wyss et al., 2022).

Environmental behaviour encompasses the actions we take that influence the natural world (Si et al., 2022). Notably, pro-environmental behaviours are those conscious actions we consciously undertake to minimize the negative impact of human activity on the environment. Examples include recycling, reducing energy consumption, using eco-friendly products, practicing sustainable transportation, and participating in conservation efforts (Jung & Kim, 2023; Thompson et al., 2009). These behaviours are often observed in individuals who hold positive environmental attitudes and a strong sense of environmental stewardship, leading them to make conscious choices that benefit the entire society (Prati et al., 2017).

Environmental behaviour of adolescents is of particular interest as they as a group would be responsible for future conservation activities and sustainability. Adolescence is a period of significant emotional, psychological, social, mental, and physical change. These transformations influence how adolescents view and interact with the world around them. The attitudes of adolescents towards the environment are strongly influenced by emotional engagement, norms, and values, which in turn shape their pro-environmental behaviours (Collado et al., 2019; Steg et al., 2011; Zhang et al., 2020). Emotional engagement refers to the development of a deep emotional connection and attachment to the environment, serving as a motivator for individuals to actively protect and care for it (Schultz et al., 2004; Taufik et al., 2015). Adolescents' behaviours and decision-making processes are significantly guided by their personal values and beliefs (Gamage et al., 2021; Sagiv et al., 2017). Particularly, values aligned with the self and transcendence play a crucial role in influencing their pro-environment behaviours, indicating their individual commitment to the preservation and protection of the environment (de Groot & Thøgersen, 2019; Frick et al., 2004; Otto, 2017 as cited in Anderson & Krettenauer, 2021). These underlying values shape their emotions, opinions, and choices concerning the environment, serving as powerful motivators (Gamage et al., 2021; Marshall et al., 2019; Sagiv et al., 2017).

Moreover, social norms, including both descriptive norms (perceptions of others' actions) and injunctive norms (perceptions of approval or disapproval), influence adolescents' environmental behaviours (Farrow et al., 2017; Keizer & Schultz, 2019). When adolescents perceive pro-environmental behaviours as widely practiced and socially valued, they are more likely to adopt them. Hence, adolescents can be considered a critical target group for environmental education and initiatives for promoting eco-behavioural change. Understanding the role of emotions and opinions in shaping their environmental behaviour is crucial for fostering positive and sustainable actions, especially as they develop their values and beliefs throughout these years.

This study aims to examine the impact of these factors on environmental behaviour, to determine the importance of targeting these components of attitudes to cultivate environmentally friendly actions among adolescents. The current study focuses on understanding the environmental awareness and behaviour of Mizo adolescents since Mizo adolescents are the future stewards of Mizoram and they possess an immense potential to be agents of positive environmental change (Chawla, 2023; Chawla & Derr, 2012; Merritt et al., 2017; Neurohr et al., 2023). This critical period in their lives, as Erikson (1968) outlined, is when they form their identities and shape their values and beliefs about the world – factors that predict their future adult behaviour (Finlay et al., 2015). Hence, the current study focuses solely on this demographic. By comparing environmental opinions, emotions, and behaviours between Mizo adolescents of different genders and educational backgrounds, the areas of strength and opportunities for improvement can be identified. Investigating the relationships between these factors will reveal how knowledge and feelings translate into action. Assessing the predictive power of environmental opinions and emotions can help identify which factors hold the most influence over adolescents' decision to act sustainably concerning the environment.

The Influences of Gender and Education

Of the many studies that have been conducted on the relationship between educational level and environmental behaviour, a number of them have shown that, regardless of culture, individuals with higher levels or extra years of education may be more inclined to adopt sustainable and environmentally friendly practices in their daily lives (Meyer, 2016; Piao & Managi, 2023; Priya & Thenmozhi, 2021; Sun et al., 2020). This can be attributed to two main factors: firstly, education equips them with a deeper understanding of environmental challenges (Meyer, 2016; Wals & Benavot, 2017). Secondly, the educational experience itself can foster a stronger sense of environmental responsibility, motivating them to act and make eco-conscious choices (Begum et al., 2022; Hadjichambis & Paraskeva-Hadjichambi, 2020).

Research has consistently shown that women tend to have more positive attitudes towards the environment compared to men (Dietz et al., 2002; Eisler et al., 2003; Milfont & Sibley, 2016; Zelezny et al., 2000). For example, studies by Milfont and Sibley (2016) and Dietz et al. (2002) found that women exhibited higher levels of environmental concern, that they were more likely to perceive environmental risks, and that they feel more responsible for protecting the environment in comparison to their male counterparts. Research by the Mintel company (2018) addressed this "eco gender gap," which may be influenced by traditional gender roles, with women often taking on household responsibilities related to tasks like cleaning, laundry, and recycling, or women having a stronger connection to nature, higher levels of empathy, and involvement in household decision-making (Christov-Moore et al., 2014; Klein & Hodges, 2001).

Research Objectives

- To compare the levels of environmental opinion, environmental emotion, and environmental behaviour among Mizo adolescents based on their gender and level of education.
- To assess the relationship between environmental opinion, environmental emotion, and environmental behaviour among Mizo adolescents.
- To investigate the predictive power of environmental opinion and environmental emotion in determining environmental behaviour among Mizo adolescents.
- To determine the main effects and interaction effects of gender and educational level on environmental opinion, environmental behaviour, and environmental emotion.

Hypotheses

- Mizo adolescents will show significant differences in the measures of environmental opinion, environmental emotion, and environmental behaviour based on their gender and level of education.
- There will be a significant association between environmental opinion, environmental emotion, and environmental behaviour among Mizo adolescents.
- Environmental opinion and environmental emotion will significantly predict environmental behaviour among Mizo adolescents.
- Gender and educational level will each exert significant main effects on the measured outcomes, that is, environmental opinion, environmental behaviour, and environmental emotion.
- There will be a significant interaction effect between gender and educational level on the measured outcomes, that is, environmental opinion, environmental behaviour, and environmental emotion.

Methodology

Design of the Study

The study employed a 2 × 3 factorial design to examine the effects of gender and educational level on pro-environmental behaviours among Mizo adolescents. The independent variables are gender (two levels: male and female) and educational level (three levels: middle school, high school, and higher secondary school). The study also featured a correlational design to assess the relationships between the dependent variables: environmental behaviour, environmental opinion, and environmental emotion. Further, hierarchical linear regression analyses were conducted to predict environmental behaviour from environmental opinion and emotion, while accounting for the potential influence of gender and educational level.

Sampling Method

Convenience sampling was employed due to limitations in resources and time constraints. Samples included middle- to higher-education students from various schools in Aizawl, Mizoram. After the process of data screening, a total of 242 responses were obtained.

Tool Used

- Environmental Attitude Scale (EAS; Uzun et al., 2019)

The Environmental Attitude Scale (EAS) consists of three sub-scales: the Environmental Behaviour Sub-Scale (EBSS), the Environmental Opinion Sub-Scale (EOSS), and the Environmental Emotion Sub-Scale (EESS). The EBSS focuses on measuring pro-environment behaviours and contains 13 items specifically designed to assess students' engagement in environmentally friendly actions. The EOSS, consisting of 11 items, captures students' opinions and beliefs regarding the environment. Finally, the EESS evaluates students' emotional responses towards the environment and includes 16 items that measure affective components such as empathy and concern. Higher scores on each sub-scale indicate a higher pro-environmental inclination towards that area.

According to the authors of the EAS, the entire scale demonstrated high internal consistency, with a coefficient of 0.94. Additionally, each individual sub-scale exhibited good reliability: the EBSS had a coefficient of 0.91, the EOSS had a coefficient of = .82, and the EESS had a coefficient of 0.94. In the current study, the EBSS had a Cronbach's alpha coefficient of 0.76, the EOSS had a coefficient of 0.66, and the EESS had a coefficient of 0.83. Thus, the EAS sub-scales were deemed fit for use among the samples.

Procedure

After obtaining permissions from the principals of various schools in Aizawl, a link to the research questionnaire (via Google Forms) was sent through WhatsApp groups to students by their respective class teachers. Participation was voluntary, and informed consent was provided that emphasized participants' rights, confidentiality, and the research purpose, following the ethical guidelines outlined by the American Psychological Association (APA Publication Ethics Code, Section 8). Detailed instructions were provided in both Mizo and English and the link to the Google Form was opened for a period of two weeks so that the participants could respond at their own convenience and at their own pace.

Statistical Analysis

The data collected via Google Forms was transferred to SPSS-16 for statistical analysis. Data screening was performed to ensure that the data satisfactorily met the assumptions for parametric testing, such as those for conducting a 2 × 3 ANOVA, Pearson's correlation analysis, and hierarchical regression analysis.

Results

Sample Characteristics

Table 5.1 shows the demographic characteristics of the participants, broken down by gender and educational level. Among the 242 participants, 38% were male and 62% were female. The table also displays the distribution of participants by education level, with 19% in middle school, 39% in high school, and 42% in higher secondary school, as well as their mean ages (by gender and education).

Table 5.2 summarizes the descriptive statistics (means, standard deviations, and sample sizes) for the environmental factors (environmental behaviour, environmental opinion, and environmental emotion) across gender (male, female) and educational level (middle school, high school, higher secondary school) categories. These statistics provide a primary understanding of the data and will serve as the basis for further analyses aimed at addressing the research objectives.

Effects of Gender and Educational Level

A 2 × 3 ANOVA was used to examine the effects of gender (male, female) and education level (middle school, high school, higher secondary school) on pro-environmental behaviour, environmental opinion, and environmental

Table 5.1 Characteristics of the samples

Demographic Variables	*Category*	N	%	*Mean Age*	*SD*
Gender	Male	91	38	16.16	1.38
	Female	151	62	15.94	1.20
Education	Middle School	45	19	14.04	0.56
	High School	95	39	15.73	0.61
	Higher Secondary School	102	42	17.18	0.48

Table 5.2 Descriptive statistics of environmental behaviour, environmental opinion, and environmental emotion scores by gender and educational category

Variables	*Gender*	*Category*	*Mean*	*SD*	*N*
Environmental Behaviour	Male	Middle School	34.58	7.05	12
		High School	30.91	6.15	33
		Higher Secondary School	33.76	6.50	46
		Male Total	32.84	6.54	91
	Female	Middle School	39.07	3.46	27
		High School	36.39	6.34	66
		Higher Secondary School	35.24	5.88	58
		Female Total	36.43	5.87	151
	Total (by level of education)	Middle School	37.69	5.20	39
		High School	34.57	6.77	99
		Higher Secondary School	34.59	6.17	104
		Total	35.08	6.36	242
Environmental Opinion	Male	Middle School	45.00	6.19	12
		High School	42.39	5.18	33
		Higher Secondary School	40.41	5.32	46
		Male Total	41.74	5.56	91
	Female	Middle School	45.26	4.39	27
		High School	43.71	6.14	66
		Higher Secondary School	42.34	6.38	58
		Female Total	43.46	6.02	151
	Total (by level of education)	Middle School	45.18	4.93	39
		High School	43.27	5.85	99
		Higher Secondary School	41.49	5.99	104
		Total	42.81	5.90	242
Environmental Emotion	Male	Middle School	61.25	6.70	12
		High School	56.73	7.37	33
		Higher Secondary School	57.87	8.92	46
		Male Total	57.90	8.16	91
	Female	Middle School	64.52	8.60	27
		High School	64.47	7.38	66
		Higher Secondary School	64.83	7.39	58
		Female Total	64.62	7.56	151
	Total (by level of education)	Middle School	63.51	8.12	39
		High School	61.89	8.21	99
		Higher Secondary School	61.75	8.78	104
		Total	62.09	8.43	242

emotion (Table 5.3). There were significant main effects of gender on pro-environmental behaviour ($F (1, 240) = 22.90, p < 0.001$) and environmental emotion ($F (1, 240) = 32.09, p < 0.001$), but not on environmental opinion ($F (1, 240) = 2.70, p = 0.10$). Females (mean (M) = 36.43, standard deviation (SD) = 5.87) scored higher on pro-environmental behaviour and reported stronger environmental emotions (M = 64.62, SD = 7.56) compared to males (M = 32.84, SD = 6.54; M = 57.90, SD = 8.16). These findings suggest that

Table 5.3 Analysis of between-subjects effects for environmental behaviour, environmental opinion, and environmental emotion variables by gender and education

Sources of Variation	*Variable*	*Type III SS*	*df*	*Mean Sq.*	*F*	*Sig.*	*Eta Squared*
Gender	Environmental Behaviour	841.60	1	841.60	22.90	0.00*	0.09
	Environmental Opinion	90.25	1	90.25	2.70	0.10	0.01
	Environmental Emotion	1958.06	1	1958.06	32.09	0.00*	0.12
Education	Environmental Behaviour	127.09	2	63.54	1.73	0.18	0.01
	Environmental Opinion	329.75	2	164.88	4.93	0.01*	0.04
	Environmental Emotion	104.66	2	52.33	0.86	0.43	0.01
Gender x Education	Environmental Behaviour	212.32	2	106.16	2.89	0.06	0.02
	Environmental Opinion	16.34	2	8.17	0.24	0.78	0.00
	Environmental Emotion	116.10	2	58.05	0.95	0.39	0.01

* Significant at the .01 level.

Table 5.4 Post-hoc Scheffe test for education's impact on environmental opinion

Dependent Variable	*I*	*J*	*Mean Difference (I-J)*	*Std. Error*	*Sig.*	*95% CI*	
						Lower	*Upper*
Environmental Opinion	Middle School	High School	1.56	1.05	0.33	-1.02	4.13
		Higher SS	3.21*	1.03	0.01	0.66	5.76
	High School	Middle School	-1.56	1.05	0.33	-4.13	1.02
		Higher SS	1.65	0.82	0.14	-0.38	3.68
	Higher SS	Middle School	-3.21*	1.03	0.01	-5.76	-0.66
		High School	-1.65	0.82	0.14	-3.68	0.38

* The mean difference is significant at the .05 level.

females tend to exhibit greater pro-environmental tendencies and emotional connection to the environment. However, both genders expressed similar levels of environmental opinion.

Education level had a significant main effect on environmental opinion ($F(2, 240) = 4.93$, $p = 0.01$) but not on pro-environmental behaviour ($F(2, 240) = 1.73$, $p = 0.18$) or environmental emotion ($F(2, 240) = 0.86$, $p = 0.43$). A post hoc analysis using Scheffe's test (Table 5.4) revealed a significant difference in environmental opinion between students in middle school and higher

Table 5.5a Relationship between environmental emotion (1), environmental opinion (2) and environmental behaviour (3) by gender

Variables	*Male*	*Female*
1 & 2	-0.00	0.15
1& 3	0.50*	0.33*
2 & 3	-0.15	0.09

* Correlation is significant at the .01 level

Table 5.5b Relationship between environmental emotion (1), environmental opinion (2) and environmental behaviour (3) by education

Variables	*Middle S*	*High S*	*Higher SS*
1 & 2	0.17	0.07	0.19
1& 3	0.55*	0.50*	0.39*
2 & 3	0.16	0.10	-0.11

* Correlation is significant at the .01 level

secondary school (mean difference = 3.21, p = 0.01, 95% CI = 0.66 to 5.76). There were no other significant pairwise comparisons (p > 0.05).

Finally, the interaction effects between gender and education were not significant for pro-environmental behaviour (F (2, 240) = 2.89), environmental opinion (F (2, 240) = 0.24, p = 0.78), and environmental emotion (F (2, 240) = 0.95, p = 0.39), meaning that gender and education together did not exert any influence on the study variables.

Relationship between Environmental Emotion, Environmental Opinion, and Environmental Behaviour

Pearson's correlation coefficients were calculated to examine the relationships between environmental emotion, opinion, and behaviour, analysed by gender and educational level (Tables 5.5a and 5.5b). The results revealed a significant moderate positive correlation between environmental emotion and environmental behaviour for both males (r = 0.50, p < .001) and females (r = 0.33; p < .001). This suggests that experiencing stronger emotional responses towards the environment is associated with a higher likelihood of engaging in pro-environmental behaviours, regardless of gender. However, no other significant correlations were found between the study variables for either gender group.

The analysis across all education levels (Table 5.5b) reveals that there is a statistically significant positive correlation between environmental emotion and pro-environmental behaviour for students in middle school (r = 0.55, p < .01), high school (r = 0.50, p < .01), and higher secondary school (r = 0.39, p < .01). These correlations were all moderate in strength.

Predicting Environmental Behaviour from Environmental Opinion and Environmental Emotion

Separate hierarchical regression analyses were conducted for males and females to explore how environmental opinion and emotion predict pro-environmental behaviour (Table 5.6a). For males, including environmental opinion in the model did not significantly improve its ability to predict pro-environmental behaviour ($\Delta R^2 = .011$, $p = .165$). However, when environmental emotion was added as a second step, the model's predictive power significantly increased ($\Delta R^2 = .255$, $p < .001$). This suggests that for males, environmental emotion is a stronger predictor of pro-environmental behaviour compared to environmental opinion.

Similarly, for females, including environmental opinion alone did not significantly improve the model ($\Delta R^2 = .001$, $p = .275$). However, adding environmental emotion as a predictor significantly enhanced the model's ability to predict pro-environmental behaviour ($\Delta R^2 = .096$, $p < .001$). This again highlights the stronger role of environmental emotion in predicting pro-environmental actions, regardless of gender.

Hierarchical regression analyses were conducted at each level of education (middle school, high school, higher secondary school) to explore how environmental opinion and emotion predict pro-environmental behaviour. The results again highlight the stronger influence of environmental emotions compared to environmental opinions.

For students in all three educational levels, including environmental opinion alone did not significantly improve the model's ability to predict pro-environmental behaviour. However, adding environmental emotion as a predictor significantly increased the model's predictive power in each case (middle school: $\Delta R^2 = .274$, $p < .001$; high school: $\Delta R^2 = .243$, $p < .001$; higher secondary school: $\Delta R^2 = .166$, $p < .001$).

Table 5.6a Hierarchical regression analysis predicting environmental behaviour from environmental opinion and environmental emotion at each level of gender

Gender	*Model*	*Variables*	*B*	*SEB*	*B*	*t*	*P*
Male	1	(Constant)	40.04	5.20		7.70	.00
		Environmental Opinion	–.17	.12	–.15	–1.40	.17
	2	(Constant)	16.79	6.18		2.72	.01
		Environmental Opinion	–.17	.11	–.15	–1.61	.11
		Environmental Emotion	.40	.07	.50	5.50	.00
Female	1	(Constant)	32.64	3.49		9.35	.00
		Environmental Opinion	.09	.08	.09	1.10	.28
	2	(Constant)	18.67	4.77		3.91	.00
		Environmental Opinion	.04	.08	.04	.51	.61
		Environmental Emotion	.25	.06	.32	4.08	.00

Note:
1. For Male students: $\Delta R^2 = .011$, $p = .165$ at Step 1; $\Delta R^2 = .255$, $p < .001$ at Step 2;
2. For Female students: $\Delta R^2 = .001$, $p = .275$ at Step 1; $\Delta R^2 = .096$, $p < .001$ at Step 2.

Table 5.6b Hierarchical regression analysis predicting environmental behaviour from environmental opinion and environmental emotion at each level of education

Education	*Model*	*Variables*	*B*	*SEB*	*β*	*t*	*P*
Middle School	1	(Constant)	28.04	7.96		3.52	0.00
		Environmental Opinion	0.19	0.18	0.16	1.08	0.28
	2	(Constant)	8.66	8.26		1.05	0.30
		Environmental Opinion	0.08	0.15	0.07	0.53	0.60
		Environmental Emotion	0.39	0.09	0.54	4.12	0.00
High School	1	(Constant)	29.86	5.10		5.86	0.00
		Environmental Opinion	0.11	0.12	0.10	0.98	0.33
	2	(Constant)	6.03	6.17		0.98	0.33
		Environmental Opinion	0.08	0.10	0.07	0.76	0.45
		Environmental Emotion	0.41	0.07	0.50	5.56	0.00
Higher Secondary School	1	(Constant)	39.39	4.31		9.14	0.00
		Environmental Opinion	−0.11	0.10	−0.11	−1.11	0.27
	2	(Constant)	24.53	5.12		4.79	0.00
		Environmental Opinion	−0.20	0.10	−0.19	−2.04	0.04
		Environmental Emotion	0.30	0.06	0.42	4.54	0.00

Note:
1. For Middle School students $\Delta R^2 = .004$, $p = .284$ at Step 1; $\Delta R^2 = .274$, $p < .001$at Step 2;
2. For High School students $\Delta R^2 = .000$, $p = .330$ at Step 1; $\Delta R^2 = .243$, $p < .001$at Step 2.
3. For Higher Secondary School students $\Delta R^2 = .270$, $p = .165$ at Step 1; $\Delta R^2 = .166$, $p < .001$ at Step 2.

This suggests that across all educational levels in Aizawl, students who experience stronger emotional connections to the environment are more likely to engage in pro-environmental behaviours. Interestingly, the strength of this association appears to weaken slightly as students progress through their education (middle school > high school > higher secondary school).

Discussion

Gender and Environmental Attitudes

The current study aligns with previous research (e.g., Milfont & Sibley, 2016) by finding that females exhibited higher pro-environmental behaviours and reported stronger environmental emotions compared to males. This supports the concept of the "eco-gender gap" (Mintel, 2018). The greater emotional connection to the environment among females may contribute to their heightened pro-environmental tendencies.

Interestingly, the study did not find a significant difference between genders in terms of environmental opinion. This suggests that while gender may influence how individuals feel and behave towards the environment, it may not have a substantial impact on their overall environmental beliefs. Unlike behaviour, which can be heavily influenced by emotions (Clore, 2011; Thaler, 2015), environmental opinions likely depend more on a combination of knowledge, logic, and personal experiences (El-Den & Sriratanaviriyakul, 2019). Therefore, the lack of gender difference in environmental opinion highlights the potential role of knowledge and reason in shaping these beliefs and that an emotional connect with the environment perhaps drives pro-environment behaviour.

Education and Environmental Opinions

The finding that middle school students had the highest environmental opinion scores initially appears to contradict the expectation of pro-environmental behaviours increasing with education level. However, some studies suggest younger adolescents might express stronger environmental concern (Pol & Castrechini, 2013; Szagun & Mesenholl, 1993). Additionally, the age range in this study is relatively narrow across educational levels. It is important to consider other influencing factors like social circles and curriculum changes that might contribute to the observed differences. Future research could explore these contextual factors for a more comprehensive understanding of the relationship between age, education, and environmental opinions in adolescents.

Emotional Connection and Pro-Environmental Behaviour

The correlation analysis revealed interesting patterns regarding environmental emotions, opinions, and behaviours. Notably, for both genders, there was no significant link between environmental opinion and emotion. This suggests that people's cognitive beliefs ("what I think" about the environment) don't necessarily translate directly into their emotional responses ("how I feel" about the environment).

However, a more consistent pattern emerged: a moderate positive correlation between environmental emotion and pro-environmental behaviour for both males and females. This aligns with previous research (Bijani, 2018; Brosch, 2021; Carmi et al., 2015) and suggests that people who experience stronger emotional connections to the environment are more likely to engage in environmentally friendly actions. This highlights the potential power of emotions in driving behaviour – emotional connection to the environment may be a stronger motivator for action than pure knowledge or beliefs.

The gender-based regression analyses further emphasise the importance of emotions. Here, environmental emotion emerged as a stronger predictor of pro-environmental behaviour compared to environmental opinion for both genders. This reinforces the notion that the emotional connection individuals feel towards the environment has a more direct impact on the choices they make.

The education-based analyses are a little more complicated. The findings suggest that pro-environmental inclinations might decrease as individuals progress through higher education and age. That is, as students progress through their educational levels, their pro-environmental tendencies may weaken. This could be due to several factors: the curriculum shifting towards a more knowledge-based approach and away from fostering emotional connection with nature, exposure to complex environmental issues that may feel overwhelming, and the development of more nuanced environmental views that might create a disconnect between emotions and actions (Hughes et al., 2019; Keith et al., 2021; Santhakumar et al., n.d. etc.).

Considering the role of emotions in pro-environment behaviour, as has been repeatedly exhibited by the results of the present study, there is reason to believe that Mizo adolescents are perhaps impacted by the natural environment with its many beautiful features and have an emotional bond to their verdant surroundings, which then may lead to behaviour aimed at conservation and sustainability. The cultural beliefs may have a role too as these beliefs and verbalizations regarding the deep connection of the Mizo with the environment are modelled by the older generation and perhaps imbibed by the adolescents.

In the context of conservation of the environment and a sustainable future, indications from the results obtained from the study are that efforts to form emotional connections with the natural world are likely to pay dividends in the long haul in terms of pro-environment behaviour. Education policy may incorporate such an approach in curricula development as well. It is emotional bonding with the environment which is most likely to lead to conservation of the environment and sustainability. Knowledge alone does not suffice for pro-environment behaviour.

Limitations

The study has some limitations. Firstly, it employed convenience sampling, which may limit the generalizability of the findings. Secondly, the data relied on self-reported measures, which can be susceptible to bias. Thirdly, the cross-sectional design precludes establishing causal relationships. Finally, incorporating qualitative data could provide richer insights into the motivations and experiences of Mizo adolescents regarding the environment.

Implications for Future Research

This study opens doors for several future research avenues. Longitudinal designs could track changes in environmental attitudes and behaviours over time. Mixed-method approaches, combining quantitative and qualitative data, could provide richer insights. Additionally, investigating contextual factors like curriculum and social influences would be valuable. Future research could also explore intervention and educational program effectiveness, comparisons across different adolescent populations, and the influence of additional variables on pro-environmental behaviours.

References

Anderson, D. J., & Krettenauer, T. (2021). Connectedness to nature and pro-environmental behaviour from early adolescence to adulthood: A comparison of urban and rural Canada. *Sustainability*, *13*(7), 3655. https://doi.org/10.3390/su13073655

Ajzen, Icek. (1991). The theory of planned behavior. *Organizational Behavior and Human Decision Processes* *50*, 179–211. https://doi.org/10.1016/0749-5978(91)90020-T

Begum, A., Liu, J., Qayum, H., & Mamdouh, A. (2022). Environmental and moral education for effective environmentalism: An ideological and philosophical approach. *International Journal of Environmental Research and Public Health*, *19*(23), 15549.

Bhattacharjee, S. (November 2017). From animism to structured beliefs: Socio-cultural changes in Lushai Hills with the advent of Christianity and British administration in the region. *International Journal of Innovative Research and Advanced Studies*, *4*, 36–40.

Bijani, M. (2018). The mediating role of environmental emotions in transition from knowledge to sustainable behaviour toward exploit groundwater resources in Iran's agriculture. *International Soil and Water Conservation Research*, *6*(2), 143–152.

Brosch, T. (2021). Affect and emotions as drivers of climate change perception and action: a review. *Current Opinion in Behavioural Sciences*, *42*, 15–21.

Carmi, N., Arnon, S., & Orion, N. (2015). Transforming environmental knowledge into Behaviour: The mediating role of environmental emotions. *The Journal of Environmental Education*, *46*(3), 183–201.

Chawla, L. (2023). Childhood experiences associated with care for the natural world: A theoretical framework for empirical results. *Children, Youth and Environments*, *17*, 144–170.

Chawla, L., & Derr, V. (2012). The development of conservation behaviors in childhood and youth. In S. D. Clayton (Ed.), *The Oxford handbook of environmental and conservation psychology* (pp. 527–555). Oxford University Press.

Chew, S. C., & Sarabia, D. (2016). Nature–culture relations: Early globalization, climate changes, and system crisis. *Sustainability*, *8*(1), 78.

Christov-Moore, L., Simpson, E. A., Coudé, G., Grigaityte, K., Iacoboni, M., & Ferrari, P. F. (2014). Empathy: gender effects in brain and Behaviour. *Neuroscience and bioBehavioural Reviews*, *46*(Pt 4), 604–627.

Clore G. L. (2011). Psychology and the rationality of emotion. *Modern Theology*, *27*(2), 325–338.

Collado, S., Staats, H., & Sancho, P. (2019). Normative influences on adolescents' self-reported pro-environmental Behaviours: The role of parents and friends. *Environment and Behaviour*, *51*(3), 288–314.

de Groot, J. I. M., & Thøgersen, J. (2019). Values and pro-environmental Behaviour. In L. Steg, A. E. van den Berg, & J. I. M. de Groot (Eds.), *Environmental psychology: An introduction* (pp. 167–178). Wiley.

Department of Environment, Forests & Climate Change. (2017). *Forest types of Mizoram*. Government of Mizoram. https://forest.mizoram.gov.in/page/forest-types

Department of Environment, Forests & Climate Change. (2020a). *Wildlife of Mizoram*. Government of Mizoram. https://forest.mizoram.gov.in/page/wildlife-of-mizoram

Department of Environment, Forests & Climate Change. (2020b). *Aizawl. Vulnerability Assessment of Forest and Biodiversity Sector due to Climate Change in Aizawl District, Mizoram*. Government of Mizoram. https://forest.mizoram.gov.in/uploads/attachments/2022/02/38c30c9470fd44955a39f9300e907655/vulnerability-assessment-report-aizawl-iora-mfd.pdf

Dietz. T., Kalof, L., & Stern, P. C. (2002). Gender, values and environmentalism. *Social Science Quarterly*, 83(1), 353–364.

Eisler, A. D., Eisler, H., & Yoshida, M. (2003). Perception of human ecology: Cross-cultural and gender comparisons. *Journal of Environmental Psychology*, 23, 89–101.

El-Den, J., & Sriratanaviriyakul, N. (2019). The Role of Opinions and Ideas as Types of Tacit Knowledge. *Procedia Computer Science*, *161*, 23–31.

Farrow, K., Grolleau, G., & Ibanez, L. (2017) Social Norms and Pro-Environmental Behaviour A Review of the Evidence. *Ecological Economics*, *140*, 1–13.

Finlay, A., Wray-Lake, L., Warren, M., & Maggs, J. L. (2015). Anticipating their future: Adolescent values for the future predict adult behaviors. *International Journal of Behavioral Development*, *39*(4), 359–367.

Forest Survey of India. (2022). *India State of Forest Report, Forest Cover 2021–2022*. Ministry of Environment, Forest and Climate Change, Government of India. https://static.pib.gov.in/WriteReadData/specificdocs/documents/2022/jan/doc2022 1207001.pdf

Global Forest Watch. (2022). *Primary Forest Loss in Mizoram, India*. Global Forest Watch Dashboard for India. https://gfw.global/49edMpf

Hadjichambis, A. C., & Paraskeva-Hadjichambi, D. (2020). Education for environmental citizenship: The pedagogical approach. In A. C. Hadjichambis, et al. (Eds.), *Conceptualizing environmental citizenship for 21st century education* (pp. 249–266). Environmental Discourses in Science Education, vol. 4. Springer.

Hmingthanzuali, H., & Pande, R. (2009). Women's indigenous knowledge and relationship with forests in Mizoram. *Asian Agri History*, *13*(2), 129–146.

Hughes, J., Rogerson, M., Barton, J., & Bragg, R. (2019). Age and connection to nature. *Frontiers in Ecology and the Environment*, *17*(5), 265–269.

Jung, Y. J., & Kim, Y. (2023). Research trends of sustainability and marketing research, 2010–2020: Topic modeling analysis. *Heliyon*, *9*(3), e14208.

Keith, R. J., Given, L. M., Martin, J. M., & Hochuli, D. F. (2021). Urban children's connections to nature and environmental behaviors differ with age and gender. *PLoS One*, *16*(7), e0255421.

Keizer, K. & Schultz, P. W. (2019). Social Norms and Pro -Environmental Behaviour. In L. Steg & J. I. M. de Groot (Eds.), *Environmental Psychology: An Introduction* (2nd ed., pp. 179–188). Wiley.

Klein, K. J. K., & Hodges, S. D. (2001). Gender differences, motivation, and empathic accuracy: When it pays to understand. *Personality and Social Psychology Bulletin*, 27, 720–730.

Lalruatpuii, E., & Ngurtinkhuma, R. K. (2023). Mizo traditional knowledge of rain and its impact on the socio-agricultural system of the society. *Mizo Studies*, *12*(3), 512–529.

Lalthangliana, B. (2005). *Culture and folklore of Mizoram*. Publications Division Ministry of Information and Broadcasting Government of India.

Leblhuber, S. K., & Vanlalhruaia, H. (2012). *Jhum* Cultivation versus the New Land Use Policy: Agrarian Change and Transformation in Mizoram. *RCC Perspectives*, *5*, 83–89.

Lodrick, D. O. (2023, April 5). *Mizoram. Encyclopedia Britannica*. https://www.britannica.com/place/Mizoram

Marshall, N. A., Thiault, L., Beeden, A., Beeden, R., Benham, C., Curnock, M. I., Diedrich, A., Gurney, G. G., Jones, L., Marshall, P. A., Nakamura, N., & Pert, P. (2019). Our environmental value orientations influence how we respond to climate change. *Frontiers in psychology*, *10*, 938.

Merritt, E. G., Rates, C., Greiner, J., Baroody, A., & Rimm-Kaufman, S. (2017). "We need trees to line the river to save our little friends": Environmental Literacy Development through Service-Learning. *Children, Youth and Environments*, 27(1), 67–90.

Meyer, A. G. (2016). Heterogeneity in the preferences and pro-environmental behavior of college students: The effects of years on campus, demographics, and external factors. *Journal of Cleaner Production*, *112*(4), 3451–3463.

Milfont, T. L., & Sibley, C. G. (2016). Empathic and social dominance orientations help explain gender differences in environmentalism: A one-year Bayesian mediation analysis. *Personality and Individual Differences*, *90*, 85–88.

Mintel. (2018). The Eco Gender Gap: 71% of Women Try to Live More Ethically, Compared to 59% of Men. https://www.mintel.com/press-centre/socialand-lifestyle/the-eco-gender-gap-71-of-women-try-to-live-more-ethically-compared-to-59-of-men

Nag, C. R. (1993). *The Mizo society in transition*. Vikas Publishing House.

Neurohr, A. L., Pasch, N., Otto, S., & Möller, A. (2023). Measuring adolescents' level of interest in nature: A promising psychological factor facilitating nature protection. *Frontiers in psychology*, *14*, 1186557.

Oskamp, S. (2000). A sustainable future for humanity? *American Psychologist*, *55*, 496–508. https://doi.org/10.1037/0003-066X.55.5.496

Piao, X., & Managi, S. (2023). The international role of education in sustainable lifestyles and economic development. *Scientific Reports*, *13*, 8733.

Pol, E., & Castrechini, A. (2013). Y 9¿Disrupción en la educación para la sostenibilidad? [Disruption in education for sustainability?]. *Revista Latinoamericana de Psicología*, *45*(3), 335–349.

Prati, G., Albanesi, C., & Pietrantoni, L. (2017). The interplay among environmental attitudes, pro-environmental Behaviour, social identity, and pro-environmental institutional climate: A longitudinal study. *Environmental Education Research*, *23*, 176–191.

Priya, B. K., & Thenmozhi, S. (2021). Study on Self-efficacy and Pro-Environmental Behavior among School Students. *The International Journal of Indian Psychology*, *9*(2), 1851–1858.

Renthlei, L. (2016). Governance and Environmental Protection in Mizoram. *Indian Journal of Public Administration*, *62*(3), 646–656.

Sagiv, L., Roccas, S., Cieciuch, J., & Schwartz, S. H. (2017). Personal values in human life. *Nature Human Behaviour*, *1*(9), 630–639.

Saitluanga, B. L. (2018). *Towards a sustainable smart city: The case of Aizawl*. ORF Special Report 73. New Delhi: ORF.

Santhakumar, V., Shephard, K., & Gupta, R. (n.d.). *Do higher levels of education make people environment-friendly? Practice Connect*, Azim Premji University. https://practiceconnect.azimpremjiuniversity.edu.in/do-higher-levels-of-education-make-people-environment-friendly/

Schultz, P. W., Shriver, C., Tabanico, J. J., & Khazian, A. M. (2004). Implicit connections with nature. *Journal of Environmental Psychology*, *24*(1), 31–42.

Shakespear, J. (1912). *The Lushai Kuki clans*. Macmillan and Co.

Singh, S. N. (1994). *Mizoram: Historical, geographical, social, economical, political and administrative*. Mittal Publications.

Steg, L., De Groot, J. I. M., Dreijerink, L., Abrahamsc, W., & Siero, F. (2011). General antecedents of personal norms, policy acceptability, and intentions: The role of values, worldviews, and environmental concern. *Society & Natural Resources*, *24*(4), 349–367.

Sun, L., Yang, S., Li, S., & Zhang, Y. (2020). Does education level affect individuals' environmentally conscious behavior? Evidence from Mainland China. *Social Behavior and Personality*, *48*(9), 1–12.

Szagun, G., & Mesenholl, E. (1993). Environmental ethics: An empirical study of West German adolescents. *The Journal of Environmental Education*, *25*(1), 37–44.

Taufik, D., Bolderdijk, J. W., and Steg, L. (2015). Acting green elicits a literal warm glow. *Nature Climate Change*, *5*(1), 37–40.

Thaler, R. H. (2015). *Misbehaving: The making of Behavioural economics* (1st edn). W.W. Norton & Company.

Thompson, R. C., Moore, C. J., vom Saal, F. S., & Swan, S. H. (2009). Plastics, the environment and human health: current consensus and future trends. *Philosophical transactions of the Royal Society of London. Series B, Biological sciences*, *364*(1526), 2153–2166.

Uzun, N., Gilbertson, K. L., Keles, O. & Ratinen, I. (2019). Environmental attitude scale for secondary school, high school and undergraduate students: Validity and reliability study. *Journal of Education in Science, Environment and Health (JESEH)*, *5*(1), 79–90.

Wals, A. E. J., & Benavot, A. (2017). Can we meet the sustainability challenges? The role of education and lifelong learning. *European Journal of Education*, *52*(4), 404–413.

Wyss, A. M., Knoch, D., & Berger, S. (2022). When and how pro-environmental attitudes turn into Behaviour: The role of costs, benefits, and self-control. *Journal of Environmental Psychology*, *79*, 101748.

Zairema. (1989). The Mizos and their religion. In K. Thanzauva (Ed.), *Towards a tribal theology: The Mizo perspective*. Mizo Theological Conference.

Zelezny, L. C., Chua, P.-P., & Aldrich, C. (2000). Elaborating on gender differences in environmentalism. *Journal of Social Issues*, *56*, 443–457.

Zhang, L., Ruiz-Menjivar, J., Luo, B., Liang, Z., & Swisher, M. E. (2020). Predicting climate change mitigation and adaptation behaviours in agricultural production: A comparison of the theory of planned behaviour and the value-belief-norm theory. *Journal of Environmental Psychology*, *68*, 101408.

Zuali, J. L. B. (2012). *Selected Mizo folktales and songs: An ecocritical reading* (PhD Thesis). Mizoram University.

Erikson, E. H. (1968). *Identity: youth and crisis*. Norton & Co.

Frick, J., Kaiser, F. G., & Wilson, M. (2004). Environmental knowledge and conservation behavior: Exploring prevalence and structure in a representative sample. *Personality and Individual Differences*, *37*(8), 1597–1613. https://doi.org/10.1016/j.paid.2004.02.015

Gamage, K. A. A., Dehideniya, D. M. S. C. P. K., & Ekanayake, S. Y. (2021). The Role of Personal Values in Learning Approaches and Student Achievements. *Behavioral Sciences (Basel, Switzerland)*, *11*(7), 102. https://doi.org/10.3390/bs11070102

JICA. (2015). https://www.jica.go.jp/Resource/english/our_work/social_environmental/id/asia/south/india/c8h0vm00009ijqki-att/c8h0vm00009pa853.pdf

Government of Mizoram. (2017). *Forest types*. Environment, Forests & Climate Change Department, Government of Mizoram, India. https://forest.mizoram.gov.in/page/forest-types

Otto, P. (2017). Nature-based environmental education of children: Environmental knowledge and connectedness to nature, together, are related to ecological behaviour. *Global Environmental Change*, *47*, 88.

Si, W., Jiang, C., & Meng, L. (2022). The relationship between environmental awareness, habitat quality, and community residents' pro-environmental behavior—Mediated effects model analysis based on social capital. *International Journal of Environmental Research and Public Health*, *19*(20), 13253. https://doi.org/10.3390/ijerph192013253

Vanlalruata, A. (1970). *Representation of mizo ethnicity in fictional narratives colonial to contemporary times*. Shodhganga@INFLIBNET: Representation of Mizo ethnicity in fictional narratives Colonial to contemporary times. http://hdl.handle.net/10603/355541

Vanlalruata, A. (2017). Rethinking Mizo Belief System. *Historical Journal Mizoram*, *XVII*, 1–16.

Part II

Embracing Environmental Intimacies

6 Bodo Community and Its Intimacy with the Environment

Mayuri Barman

The role of the environment in securing subsistence and in sustainable development is crucial. Sustainable development leads to more and more perfection stands in need of environmental stewardship, which implies a sense of mutual care to be spearheaded by human beings themselves. What is required is not the attitude of consumerism and reckless use of natural resources, domination over and exploitation of nature, but responsible use, equitable distribution, and preservation of the splendors of nature. In this respect, it can be stated that nature has been worshipped in India since the ancient times to form a harmonious relationship between itself and other living beings. Man worships the earth, the mountains, the rivers, the oceans, the trees, the plants, and the birds. All that is required is respect for the natural world, which keeps us alive and feeds us.

In this beautiful journey of man's development, nature has proven to be the true companion of man; his relationship with nature has been symbiotic and friendly. But with the advancement of science and technology, the relation between man and nature and man and other life forms in nature changed from symbiotic to predatory as man was inclined towards maximum use of natural resources. Thus, in today's world a range of environmental problems such as global warming, deforestation, reduction in biodiversity, and others have become a reality. Future development appears to be limited and constrained by the growing population and global climate change. Therefore, for the future development of society humans need to seek growth in a sustainable manner.

At a time when ecological degradation and deforestation have been taking place at an alarming rate throughout the globe, in India thousands of pockets of natural vegetation scattered throughout the country are preserved almost in pristine condition. Almost all the religious communities and tribal societies consider some plants holy in origin and essential in religious functions. The indigenous people all over the world have accumulated knowledge over time, which in traditional set-ups allow them to live in balance with their environment and helps them to adjust with the natural resources for their livelihood. Further, such knowledge systems – conceptualized as indigenous knowledge systems – are handed down through oral tradition as well as through various sacred rituals, cultural practices, and beliefs in which they remain embodied.

DOI: 10.4324/9781003504870-8

Figure 6.1 A tree showing sign of worship by man. (Photograph by the author).

These together are an integral part of the psychological make-up of every individual of the indigenous communities.

Plant worship, as part of nature worship, is generally believed to have begun in the initial stage of human society. Many ethnic people have the tradition of worshiping different trees on different occasions. While moving around in their environment, if they come across any sacred tree, they stop and tie a thread around the trunk of the tree or put flags near the tree to show their reverence.

Culture and Nature

Plants are used in many ways including worship and betterment of human life. In every human society worship is performed with traditional rituals for well-being. Many tribal communities preserve this tradition through folklore and worship their deities right from the occasion of birth to mourning death. They perform specific worship with precise offerings. Tribal folklore is rich in magico-religious beliefs. They believe that some Gods and deities reside on the trees in the forest. If they do not show respect to them their full clan will be destroyed. Therefore, they preserve the plants which they regard as sacred for social, cultural and religious purposes. Their festivals, rituals and other cultural aspects are closely associated with the surrounding vegetation preserved on religious ground. The extra care exhibited by the communities has significantly contributed to preserving the religious plants and thereby conserving a whole range of biodiversity. There is an inextricable link between indigenous people and biological diversity. All over the world the indigenous people have protected the biodiversity with which they have a symbiotic relationship.

It is an undeniable fact that the knowledge of indigenous people is invaluable in the present-day context of biological diversity conservation and its sustainable utilization. Plants have a special role in religious and social ceremonies of every rural society. Various religious and supernatural beliefs and folklore help in the prevention of destruction of plants. There are several examples of tree worship tradition in many parts of the world under all religions and beliefs.

Assam and Its Tribes

Assam, situated at the foothills of the Eastern Himalayas, is the largest state in North East India and lies in the middle reach of the rivers Brahmaputra and Barak. The land has uneven topography, falls of hills, plains and rivers. The Brahmaputra River flows through Assam from east to west over a length of approximately 650 km.

The entire North Eastern region comprises of states with rich floristic diversity and is also inhabited by the largest number of tribes who lead an intricate life totally dependent on the forest plants. More than 25 communities mainly depending on plant resources for their day-to-day life inhabit the different parts of Assam. They have close association with and good knowledge about plant resources in their surroundings which form an integral part of their material and spiritual culture.

The tribal people are very attached to the forest and interact greatly with such an ecosystem. Their economy, religion and social organization revolve around the forest eco-system. Generally, forests and tribal people are inseparable and the existence and development of one depends on the other. Therefore, it is said that "the tribal communities in India largely occupy forest regions where for a long period in their history, they have lived in isolation but in harmony with nature."

They worship nature and natural products. They call the forest '*Adivi thalli*,' the hill or hillock as '*Konda Devata*'. They believe that the forest nourishes them and provides food security for their livelihood. They believe that their Gods and Goddesses take abode in forests, and they consider hills and forests as their sacred complexes. They propitiate the forest and hill Gods and Goddesses abundantly with a view to get protection from wild animals, snakes and natural calamities. Some of the objects or beliefs, which are environmentally prominent, are given in the following.

Sacred Trees

They consider the following trees sacred in nature, which they do not cut either for firewood or for house construction, agricultural and other domestic purposes.

These trees are Ravi (*Ficus religiosa*), Jammi (*Prospis Cinararia*), Ippa (*Madhuka Longfolia*), Marri (*Ficus Benghalensis*), Neem (*Margosa Indika*), Doudera (*Butia fromdosa*), Jalli (*Arachis futicosa*), Sal (*Sarji*), Banyan (*Bara*), Salap (*Salapa*), Blackberry (*Jammu Koli*), and Jackfruit (*Panasa*).

Religious Beliefs

Religion plays a significant role in sustainable development. According to Tyler (1871), "Religion is essentially bound in souls or spirits of a personal kind, as the causes of natural events".

Animism

Animism is the belief in supernatural powers that organize and animate the material universe. Animism regards each spirit to be bound to its object. Thus, man's spirit is subject to nature. Among the sacred beliefs, the faith in spirit is most common with the tribals. This animistic belief is rather a universal feature of the religion the tribals have. For them all spots and places are holy as they are the seats of spirits. Animals, plants, trees, ponds, rivers, hills or mountains are all abodes of spirits. Thus, it gradually develops into spiritism.

Totemism

Apart from nature the indigenous people have also congregated themselves with the animals, plants in the form of 'Totem'. "Totemism is a system of belief in which humans are said to have kinship, or a mystical relationship with a spirit-being, such as an animal or plant" (https://www.britannica.com). 'Totem' was their God, and they made offerings to please their totem for the benefit of the tribe. Thus, the struggle for self-preservation took the form of clan preservation.

Totemic Belief

The existence of totemic belief also plays a significant role in maintaining biodiversity. Several tribes claim to have mysterious or mischievous relationships with some physical objects, namely plants, animals and so on. According to Frazer (1910), "A totem is a class of material objects with a savage regard with superstitious respect, believing that there exists between it and every member of the clan an intimate and altogether special relation. The connection between a person and his totem is mutually beneficent, the totem protects the man, and the man shows his respect for the totem in various ways, by not killing it if it be an animal, and not cutting or gathering it if it be plant."

The elements of totemic clans are also present in the greater traditions. The names of some of the Rishis providing clan (*gotra*) names, such as Kashyapa

(Tortoise), Gotam (Cow), Vata (Calf), Sunaka (Dog), Bharadwaja (a kind of bird) and others.

The following totemic clans are also found among the tribes – (1) Ontala (Snake), (2) Killo (Tiger), (3) Gollori (Monkey), (4) Kimudu (Bear), (5) Matsya (Fish), and (8) Chilli (Goat). These totemic objects are considered as sacred and killing or eating of the flesh of these is a taboo. Thus, the natural objects of flora and fauna have religious, economic and social significance which are significantly attached to the lifestyle of these tribal people. As the Musahari tribes of Bodo, Khamti tribes of Arunachal believe that their forefathers were tigers, therefore they never kill tigers. Lotha and Bangia tribes of Nagaland believed that their forefathers were snakes (*Ajagar*). Thus, it is found that many tribes believed that their forefathers were either flora or fauna. So, it is their duty and culture to conserve nature and the environment.

Although certain practices themselves, such as plant worship, or plants used in religious festivals, may seem dated in the modern world view, their essence must form the foundation of current efforts at conservation and sustainable development.

The Tribes of Assam

Karbi

Racially the Karbis belong to the Mongoloid group and linguistically to the Tibeto-Burman group. Based on their pattern of distribution or habitation, the Karbi are divided into three groups: Chinthong, Ronghang and Amri. The traditional religion of the Karbis is animistic, however in modern times, a small percentage of the Karbis have adopted the Vaishnava and other Hindu faiths and a few follow the tenets of Christianity. They are wholly dependent on agriculture for their livelihood. '*Rongker*' and the '*Hacha-Kekan*' are the two main festivals of the Karbis. The former is celebrated before the Jhum cultivation starts in winter and again in the summer, just prior to the planting of rice. The '*Hacha-Kekan*' is celebrated at the end of the planting season and the dance performed at the harvesting festival is lively.

The Missing (Mishing)

The Mishing, an Indo-Mongoloid group formerly referred to as Miris, are the second largest ethnic group in Assam. The Mishing tribe moved to Arunachal Pradesh during the eigth century, coming from northern territories, through the world's highest lake, Manasa Sarovar. They later migrated down the course of tributaries feeding into the Brahmaputra to the Assam plains in search of a more economically viable life on the fertile riverbanks. The traditional religious beliefs of the Mishing are animistic. Mishing women are renowned for their weaving art, crafting shawls and blankets. The two chief traditional festivals observed by the Mishing are the '*Ali-Ai-Ligang*' and the '*Porag*'. *Ali-Ai-Ligang*, the most colorful, is the spring festival and *Porag* is the post-harvest festival.

Sonowal Kacharis

The Sonowal Kacharis are one of the royal dynasties of the North East and are a branch of the great Bodo-Kacharis, a major tribal community in Assam. They are predominantly inhabitants of the Dhemaji, Lakhimpur, Tinsukia and Dibrugarh districts of Assam. They belong to the Mongoloid race of people. A portion of Sonowal Kacharis still follow the ancient religion of animism, where nature is seen to be alive with invisible forces, minor deities, and spirits, with which priests and medicine men mediate.

Deori

The Deori live along the Brahmaputra River, primarily located in the states of Assam, Nagaland, and Arunachal Pradesh. They speak a Tibeto-Burmese language which is also called Deori. One of the noted social peculiarities of the Deori is that entire extended family units live under one roof, until sometimes more than 100 people live in the same house. They observe two major festivals during the year, '*Isaku Bou*' in the month of April and '*Magiyo Bisu*' in the month of January; both relate to agricultural activities and are observed before starting the agricultural operations in the fields. '*Joydam*' festival is another major festival observed every year in the month of February.

Rabha

The Rabhas are one of the nine scheduled tribes in the plains districts of Assam and are widely scattered but mostly concentrated in Goalpara, Kamrup and Darrang districts. The Rabha are traditionally animist. However, today their faith involves some Hindu and a few animistic rituals. The ritual practices among forest Rabhas who still live in the forest and the Rabhas that live in the villages as cultivators are found to be different in ritual practices. The forest Rabhas follow traditional animistic practices tinged with some rituals of mainstream Hinduism. On the other hand, village Rabhas have merged with local Hindus as far as their religious practices are concerned. In the past, the Rabhas practiced shifting or Jhum cultivation.

Dimasa

The Dimasa are one of the oldest inhabitants of North East India and one of the many Kachari tribes, living mostly in Dima Hasao in Assam. Dimasa means 'children of the Big River.' Dimasa tend to live on riverbanks and next to streams. Agriculture is the principal occupation and the main source of livelihood for the Dimasa Kacharis. Among the festivals of the Dimasa,

'*Bushu*' is the most joyous and important community festival celebrated every year on 27th January, when all work in the field is completed. '*Bushu*' is an occasion for relaxation from hard work and is termed as harvesting festival or a festival of rejoicing and merrymaking.

Tiwa

Tiwa is an ethnic group inhabiting the states of Assam and Meghalaya. They were known as Lalungs in the Assamese Colonial literature and in the Constitution of India, though members of the group prefer to call themselves Tiwa. A striking peculiarity of the Tiwa is their division into two sub-groups, hill Tiwa and plain Tiwa, displaying their own cultural features. The hill Tiwas live in the Westernmost areas of Karbi-Anglong district in Assam as well as in the North Eastern corner of Ri-Bhoi district, Meghalaya. They speak the Tibeto-Burman language of the Bodo-Garo group. The other half have been converted to Christianity since the 1950s. Plain Tiwas live on the flat lands of the Southern bank of the Brahmaputra Valley, mostly in Morigaon, Nagaon and Kamrup districts.

Tiwa people are closely associated with the principality of '*Gobha*' – their king whose territory covers the Tiwa cultural realm. The *Gobha* king has been mentioned in Assamese colonial literature since the early eighteenth century as an important mediator in trade between Ahom dominated plains, that is, Assam and the Jaintia Kingdom. The historical role of *Gobha* and the Tiwas as mediators between plains and hills in Central Assam is enacted every year during an ancient fair, the Jonbeel mela.

Tai-Phake

Tai Phake forms a Thai-speaking ethnic group based in Dibrugarh and Tinsukia districts in Assam and some of them can be met in Arunachal Pradesh. The Tai Phake were believed to have migrated from the Ahom Kingdom of Myanmar and settled in the rich south bank of the Dhing river in Assam in the eighteenth century. The main occupations of the Tai Phake people are agriculture and fishing.

Tea-Tribe

The Tea tribes, also called Adivasi, are the tribal people who were brought into Assam by the British Colonial planters as indentured laborer from the Chota Nagpur Plateau region of India during the British Raj. They live in almost every district of Assam, but their density varies according to the number of tea plantations in different regions of Assam.

The largest tribal group in Assam is the Bodo Kacharies who belong to the great Bodo group of the Indo-Mongoloid family and are believed to have their origins in Tibet and China. Being the major tribe in Assam, and with a presence which is scattered across a large part of Assam, their beliefs and rituals are discussed in detail in the following section.

The Bodo

The Bodo is the largest ethno linguistic group in the Assam state of India. They are a part of the greater Bodo Kachari family of ethno linguistic groups and are spread across North Eastern India. The Bodo along with other cognate groups of Bodo Kachari people are prehistoric settlers who are believed to have migrated at least 3000 years ago. The Bodo tribe is known to be the earliest immigrants in Assam. The ethnic community is mainly concentrated in the Brahmaputra valley. It is the largest minority group of the state, and it is believed that this tribe arrived in Assam from Tibet through Bhutanese passes. Their rich culture incorporates elements like dancing, singing, and so on, which reflects the fact that they have many religious practices and beliefs, among which Bathouism has special importance.

The Bodos believe in different Gods and Goddesses. They believe in five traditional Gods, such as *Ailong* (God of earth), *Agrang* (God of water), *Khoila* (God of air), *Sanjaborle* (God of fire or light) and *Rajkumbre* (God of sky). The popular deities are grouped into two classes: (1) Household deities (*na-ni-madai*) and (2) Village deities (*gami-ni madai*). The household deities are worshipped inside the house or in the house compound, whereas the village deities are worshipped outside the houses, usually near the sacred grove of trees or bamboos, often to be seen some fifteen or twenty yards from the village and known as the *thansati*.

Bathou is the most important God. He is never represented in idol form, but is well in evidence through his living symbol, the '*Siju*' plant (*Euphorbia Splendens*), which is often to be seen in the Bodo homestead surrounded by a circular fence of split bamboo. In this respect it has a similarity with the Hindu religion because some of the Hindu people worship the *Tulsi* plant in front of the house. Writers who take *Bathou* as a Forest Deity cannot satisfactorily explain the religious and philosophical significance of the *Siju* plant as impressed upon the Bodos who have worshipped it since time immemorial.

Trees and plants are the direct objects of invocation in the Rig Veda (V-41.8) and the Brahminic law enjoins upon the faithful to sacrifice an offering to the great Gods, 'to the waters' and 'to the trees' as is the case in the house ritual. It is true, however, that the greater veneration due to some trees and plants has a special reason. For example, *Tulsi, Ashok* tree, *Siju* plant and others have medicinal values. On the other hand, a pair of trees, namely a banyan tree (Ficus bengalesis) as bride and Peepal (Ficus religion) on parlease (Ficus venesa) as groom are closely planted together as sacred plants. This is termed as married couple plants known as *Bar-Kaina* (Groom-Bride). The sprouting leaves of these trees are very essential for many pujas in the Hindu religion.

Figure 6.2 Photograph of the *Siju* plant by the author.

Figure 6.3 Photograph of another *Siju* plant by the author.

Figure 6.4 Photograph by the author of a '*Bathou Thansali*' (altar of God Bathou).

Figure 6.5 Photograph by the author of an altar to some deity in the forest.

Figure 6.6 Photograph of *Bar-Kaina* by the author.

The use of *Siju* plant can cure people from some chronic diseases, for example, through the preparation of poultices. Again, the '*Bathou*' is associated with the numeral five in its quality and function. The *Eupherbia splendens* is characterized by five ridges and five depressions which correspond to five elements of creation and represent truth, knowledge, power, beauty and goodness. This plant symbolizes '*Bathou*'. The Bodos believe that this plant was created first among the vegetations in this universe and men took shelter under its shadow. Hence, they accept it as a symbol of the Supreme God '*Bathoubrai*'.

Kherai is a spiritual and largest ceremonial performance which manifests collective hope and aspiration for the community. The Bodos are generally dependent on agriculture, so before and after cultivation they perform *Kherai* puja and propitiate different Gods and Goddesses for success in cultivation. In the *Kherai* puja, along with the deities of the Bodos, the Gods of different leading religions of the world are also worshipped according to the respective customs. Those abodes are *Rungsar, Mitha, Hashing* or *Hajing*, equivalent to other religions known as *Swargo, Mortya and Patal.*

Dances of the Bodo

Bagurumba

Bagurumba dance is an important dance of the Bodos of Assam. This dance is usually performed during Baisagu festival. The Bagurunba dance is called butterfly dance. It resembles the movement of birds and butterflies. It is instituted to propitiate ', the supreme deity of the Bodos.

Satrali

Satrali dance is known as sword welding dance. It is performed to show fighting in a war. While dancing the *Doudini* holds swords in hand and in her mouth and dances to the tune of *Kham*, *Siphung* and *Taal*. It is performed to praise the deity of fire or sun known as *Agrang*.

Gorai

Gorai dance is known as dance of horse riding as '*Gorai*' means 'horse'. This dance is presented to mother earth during *Kherai* puja. The *doudini* here symbolizes horse-riding and dances around the *kherai* arena which represents the physical and will power of human beings.

It is conjectured that the supreme deity whom people worship and pray to is supposed to be a single one. Yet people presumed that the supreme deity must have been supported and helped in different fields by those who are known as messengers. Accordingly, the supreme power whom we supposed as God distributes the work and responsibilities for smooth administration. In the *Kherai* worship performed in the traditional style with elaborate ritual, sacrifice, and shamanic dance of the *Doudini*, eighteen male and female deities in pair are invoked. The worship of pluralistic deities is present in Hinduism and other religions too. The traditional religion of the Bodo has some resemblances with the ancient practice of Hindu religion. Even the *Bathoubrai* (the Supreme God) is understood as the embodiment of the trinity in one.

Archaeological evidence indicates that Bodo tribes had the influence of Chinese culture of the East and the Egyptian and Babylonian culture of the West probably due to their trade relations with the people of those countries. In religion, in their original homeland, people were worshippers of the symbols of vegetation sprouting in the form of a twig of *Euphorbia splendens* that grows naturally with all its stems sprouting vertically upwards with five ridges and five depressions. It is said that the *Siju* plant has medicinal value on its every knot of the stem, row of thorns, sprouting leaves and so on. Hence, '*Siju*' plant is a symbol of *Bathou* and worshipped as such.

The Psychological Effects of Cultural Practices

There is numerous research that implies that religious and cultural practices that share close relationship with the environment contribute to one's mental health. It is believed that cultural practices fill up the void that may arise due to existential angst that one may face during the journey of life and find order in the chaos around and within oneself. A study by Shariff and Norenzayan (2007) stated that religious practices encourage people to be more charitable by promoting belief in supernatural agents. This may therefore be the reason why people may see religious and cultural practices in the same light. Religious practice may become a part of the culture as evident in the practices of the Bodo tribe where the surroundings signify the faith one bestows on the existence of the supernatural.

Conclusion

The preceding discussion reveals that from the beginning of human society, each tribe develops its own ways of utilizing and managing the community and natural resources. As the indigenous people are living at the mercy of nature, they have developed various conservative measures for the security of their own community. Depending on environment and cultural values, the procedures of utilization vary from community to community, but it is a universal fact that human beings were always aware of balancing the ecosystem and maintaining harmony with nature.

It is apparent that the knowledge system or precautions the indigenous people emphasize on is entirely meant for ensuring larger social security. Various socio-cultural beliefs and practices have emerged during the development of human society. The sacred conservation practices, superstitions, religious taboos and prohibitions no doubt play an important role in protecting the natural resources and thereby contribute to managing the ecosystem. Thus, it is the responsibility of social scientists and anthropologists to explore the indigenous processes because there rests the basic philosophy for the survival mechanisms of society and nature.

Finally, it can be opined that the importance of environmental ethics in present-day society is indispensable. As far as environmental ethics is concerned, ecological consciousness leads towards the welfare of all in the environment and nature. As it is said – "*Sarve Sukhina santu sarve santu niramayah sarve bhadrani pashyantu ma kaseidukhabhaghavet*" [May all be happy. May all be free from disease. May all realize what is good. May none be subject to Misery].

Utilitarianism maintains that the balance of pleasure and pain should be taken equally into consideration. In this regard the utilitarianism of J.S. Mill and Bentham has prescribed happiness for all – "*Greatest happiness of the greatest number*". Therefore, human beings, along with the other forms of life, are a part of the food chain and are closely associated with each other; together they form the natural environment.

The intimacy of the tribes, and the Bodo in particular, with nature is evident from their religious practices and their festivals along with the associated dances. In almost every aspect of their life and culture, the natural environment along with its elements is intertwined. The Bodo organizations have made attempts to ensure that these religious and cultural practices are carried on by later generations. It is this passing on of the environmental intimacies which may promote conservation and respect for the Earth and its elements.

Bibliography

Anderson, J. D. (2002). *Introduction to Kachari tales and Rhymes*. Shillong Secretariat Press.

Arumugam, E. (2008). *Principles of environmental ethics*. Sarup & Sons.

Baruah, I. (1998). *Socio cultural and environmental factors of health and economy*. Indian Society for Ecology, University of Delhi.

Bhatt, S. R. (2003). *Applied philosophy and professional ethics*. ICPR.

Bordoloi, B. K., & Athparia, R. K. (2003). *People of India, Assam*. Vol. XV. Anthropological Survey of India. Seagull Books.

Bordoloi, B. N. (1987). *Tribes of Assam, Part-I*. Tribal Research Institute, Assam.

Brahma, K. (1989). *Aspects of social customs of the Bodos*. Bina Library, Guwahati.

Chaudhuri, B. (1986). *Tribal health: Socio-cultural dimensions*. Inter-India Publications.

Das, B. C., & Sharma, C. (2012). *Moral philosophy: Contemporary perspectives*. EBH Publishers (India), Guwahati.

Dasgupta, S., & Sarkar, A. (2000). *Ethno ecology of Indian tribes*. Rawat Publications.

Deogaonkar, S. G. (2006) *Environmental preservation and tribal populations*. Vol. XXIX. Tribal Research Bulletin.

Doshi, S. L. (1995). *Anthropology of food and nutrition*. Rawat Publications.

Eliade, M. (1949). *Patterns in comparative religion*. Sheed & Ward.

Eliot, R. (1995). *Environmental ethics*. Oxford Readings in Philosophy.

Endle, S. (2000). *The Kacharis (Bodo)*. Low Price Publications.

Frazer, J. G. (1910). *Totemism and exogamy: A treatise on certain early forms of superstition and society*. Vol. 1. Macmillan and Co. Ltd.

Freud, S. (1919). *Totem and taboo*. Penguin Books Ltd.

Gohain, B. K. (1993). *The Hill Lalungs*. Anundoram Borooah Institute of Language, Art and Culture, Assam.

Ghosh, Subodh Kr. (1967). *Handbook of philosophy of religion*, Indibook Distributing Co.

Hashi, S. S. (1974). *Encyclopedia of Indian tribes of the tribal world in transition*. Vol. I. Anmol Publications Pvt. Ltd.

Kolay, S. (2001). *Ecology, economy and tribal development*. Mohit Publication.

Medhi, B. K. (2009). *Tribes of North East India: Issues and challenges*. Omsons Publications.

Pandey, G. N. (1997). *Environmental management*. Vikash Publishing House Pvt. Ltd.

Reddy, K. V. (2002). *Cultural ecology of Indian Tribes*. Raj Publications.

Rolston, H. (2003). *Environmental ethics*. Oxford Blackwell Publishing.

Rajendran, S. M., & Agarwal, S. C. (2007). Medicinal plants: Conservation through sacred forest by Ethnic Tribals of Virudhunagar District, Tamil Nadu. *Indian Journal of Traditional Knowledge*, 6(2), 328–333.

Shariff, A. F., & Norenzayan, A. (2007). God is watching you: Priming god concepts increases prosocial behavior in an anonymous economic game. *Psychological Science*, *18*(9), 803–809. https://doi.org/10.1111/j.1467-9280.2007.01983.x

Tyler, E. B. (1871). *Primitive culture*. Vol. 1. John Murray.

7 The Folk Culture and Ritual Practices of the Karbi

Bamonkiri Rongpi

Culture and society are intertwined and rely on one another. Society can be viewed as a process, a sequence of human interactions in which each person reacts to the stimulus of the other. Without such interaction no social life is possible as communication is the foundation to all social life. All human societies can be seen as an outward expression of their members' social interactions. The norms, beliefs, and practices associated with the social structure determine the cultural elements of people and these interactions eventually shape the social structure. A society is an organized, interactive group of people who adhere to a particular way of life. Therefore, society is made up of people and how they behave is their culture (Herskovitz, 1955).

The term 'folk culture' includes unity, tradition, and folklore that is a collective combination of myths, legends, tales, proverbs, riddles, ballads, folk song, folk dance, folk music, folk carvings, and so on. As language is the vehicle of culture, folk culture has been carried over generations, orally, since time immemorial.

Environment and life have a very close relationship. Every change a living being undergoes entails some changes in the relation between the organism and the environment and vice versa. The environment serves as a place of habitat. The means of livelihood and the cultural aspect are directly correlated with the environment. In the cultural environment of a society, culture has its own implication to the ways of life in that community.

India is a multi-religious and multi-ethnic nation. In addition to being multi-ethnic, the cultural variations are diverse in kind. With the geographical landscape's diversity, culture has evolved to differ from one place to the next. The tribal culture is significant within the framework of India's multiculturalism. Most Indian tribes are in forested, hilly, and geographically remote areas. The daily activities of the tribal people are strongly tied to their immediate environment. Tribal culture has a very rich history and is fed by nature. The tribal people's cultural traditions are largely dependent on folklore that are informal, oral, illiterate, and unreflective. Their cultural history, as well as the characteristics and features that define it, belong to the entire community and not just to the individual members. The tribal people and their folk traditions are a composite whole and this is one of their most distinctive cultural traits.

DOI: 10.4324/9781003504870-9

Folk songs, verse, and music go together, and the musical aspect is reflected more in its practice. The simple catchy folk tunes conveying the feelings and sentiments of their ancestors, their lives and thoughts, their activities and achievements, their morals and discipline are of great value (Vidyarthi and Rai, 1976). The term 'folk culture' includes unity, tradition, community dependence and collective awareness and consciousness of the concerned people. This term was first coined by W. J. Thomas in 1846 and it is recognized universally. C.F. Potter has defined folk culture as "a lively fossil which refuses to die" (cited in Deogaonkar and Deogaonkar, 2003).

Many people share and find value in social customs, rituals, and festive occasions that structure the life of communities and groups. They are significant because, whether performed in public or privately, they are strongly associated with significant events and reaffirm the identity of those who practice them as a group or society. Rituals and customs can be used to commemorate life phases, agricultural calendar events, and the change of the seasons. They are strongly related to a community's outlook on the world and understanding of its own past and memories.

The Karbi are the early settlers of Karbi Anglong. They have their own folk culture which is associated with different ritualistic practices. They have a very rich cultural tradition, material culture, and folklore. The methods of transmission of folklore are purely based on listening to the songs sung by the singer, stories told by grandmothers or elders, and recitation of verses during *Se-Karkli* and other ceremonies. The Karbi cultural tradition is purely based on oral transmission.

Among the Karbi several rituals can be noticed. Most of their rituals are performed at the community level in the village and all the villagers collectively participate in them. The concerned people are intimately dependent upon nature for their sustenance. Such performances are not merely for their recreation but denote their devotion to supernatural powers and agencies. Their entire worship is focused and related to the issue of health, safety, security, protection, and betterment of the entire community.

Different natural resources play a significant role in the livelihood and cultural tradition of the Karbi people. They have their own concept of ethno-ecology and indigenous knowledge sustaining their age-old cultural heritage. Different aspects of material culture reflect their ethno-ecological perspective and indigenous knowledge base related to preservation and continuation of traditions over generations. However, the folk culture of the Karbi is facing a serious challenge and may be gradually on the way to obliteration.

The Karbi People

Karbi Anglong, one of the greenest districts of Assam, has its administrative headquarters at Diphu town. The prefix of the name of the district is taken from the name of the community *Karbi*, which is the numerically dominant community who are indigenous to not only the district but also the state.

The word *Anglong* means hills and mountain. Thus, the name literally means hills of the Karbi.

In the ancient times, the hill tribes of North East India were not a part of India; neither was it a part of Assam before the British colonized the region. Since the recommendation of the Simon Commission, the hill areas of Karbi Anglong district were granted a different status under section 92 of the Government of India Act 1935. The Mikir Hills, which was the old name of Karbi Anglong, was placed under the 'Partially Excluded Areas'. The then government of Assam had no jurisdiction over the excluded areas, and it was administered under the special power of the governor.

The district enjoys autonomy under the provision of the sixth schedule of the Constitution of India. It is the largest district of the state with an area of 10,434 sq km and covers 13.3 percent of the total land area of the state. The district was further divided into two parts in 2016, Karbi Anglong district and West Karbi Anglong district. The district is bordered by Golaghat and Nagaon district in the north, Dima Hasao district and the state of Nagaland in the south, Golaghat district in the east, and Morigaon district and the state of Meghalaya in the west.

The plateaus of the district are an extension of the Indian Plate (The Peninsular Block) in the North East of India. The major rivers that flow through the district are Kopili, Amreng, Nambor, Borpani, Deopani, Chomna, Koliani, Patradisa, Dikhru, Longnit, and others. Variation in the topography makes the hilly areas of the district experience a different type of climate than the plain areas of the district. The winter season starts in October and continues till February. The district receives abundant rainfall during monsoon in summer, from June to September. The district has a pleasant climate throughout the year while the monsoon season has plenty of rainfall. The average rainfall of the district is about 2416 mm.

According to Jones Ingti Kathar, the Karbi after their separation from Asia reached the Himalayan region of Tibet. While they were traipsing, they came across a swarm of bees, birds, and animals. All of these were logged and listed in the form of a dance named as '*Banjar-Kekur*'. The Himalayas covered with ice made the tribe perceive that these summits are the sanctuary of the Gods, namely *Hemphu, Mukrang, Rasinja, Bamon, Sojon*, and so on. Cumulative prayer was conducted in Tibet which is known as '*Peng Karkli*' today. The stay of the Karbi in Tibet is authenticated by the mention of '*Lhasa*' in the song they sing during the '*Chomkan festival*'. *Chomkan festival* is the festival dedicated to the dead. The festival lasts for four days and four nights continuously. This festival does not invite anyone formally but everyone who wishes to be present at the ceremony is welcome. This festival is a requisite for the Karbi. The word '*Arleng*' which is the alternative for Karbi, was probably named during their stay in the hilly region of Tibet. The next lay-off of the Karbi was in Myanmar (Burma).

The finding or discovery of paddy was the first and foremost recorded activity of the Karbi in Burma. The story regarding this discovery is found in the song '*Lokhi Keplang*' or '*Porom Alun*'. This song is sung relating it with Jhum cultivation. According to B.I Kathar, paddy was discovered by the tribe in Tibet, however threadbare observation points to Burma. K.C. Tensuba (1991) in his book *Genesis of Indian Tribes* stated that tribal groups coming from the north of the mountains moved down to learn the art of cultivation. The later ones or the weaker groups of people were pushed into the mountains to learn only '*slash or burn*' means of rice growing.

The celebration of a marriage ceremony at *Muring-Murong* village was the next activity of the Karbi during their stay in Burma. Unfortunately, the day the marriage ceremony was fixed Thai people (Shantai) came to fight a battle against the Karbi on the issue of pollution of the upper stream by the Karbi. The matter was resolved; however, a risk of battle continued towards the Karbi from the other side. Hence, the Karbi left the village of *Muring-Murong* and reached the Arakan area and the Patkai hill range. Some places like Salu, Mukindon, Masum Longle, and Langherang and Ningthi river were known to them. The Karbi took three different routes while coming to the land of Assam. The first route they took was through the Patkai hill ranges. Second, through Manipur from Chindwin. Third, from Irrawind through Arakan. It has been noticed that it was extremely difficult to confirm the exact places and affinities of the heterogeneous congeries of the people who inhabited the mountainous region between India and Burma (Myanmar) ever since the races have been studied and understood. This was observed or mentioned by Robinson in 1841 and 1849, by Steward in 1855, and by Dammant in 1879. There was also an attempt to bring the Karbi into relation with the Boro of Kachari stock, which was shown in the census in 1881. However, it was seen that the time was ascertained more regarding their neighbors and a final judgement could not be arrived at (cited in Teron, 2008).

Folk Culture of the Karbi

Like any other tribe, the Karbi have their own culture, language, folklore, traditions, and religious belief. Various folk cultural practices such as riddles, legends, folk songs, lament songs, folk dances, and so on can be seen in their society. In every step of the rituals practiced in Karbi society, narration of every activity right from the beginning till now is done through folk song. The oral traditions prevalent among the Karbi for generations in the form of songs, dances, music, and folktales reflect various dimensions of their community, including their idea about the origin of the Karbi. Some of the traditional songs prevalent among the Karbi are found to be very interesting and provide information regarding their society and culture. There were times when the Karbi ancestor could speak endless verses, sing songs, and perform dances which are associated with religion, marriage, death, birth, and many other folk cultural performances. In ancient days, the Karbi ancestors were very fluent in the language that they used to express their views, religion, and origin and to communicate with their surroundings and descendants.

The knowledge of verses, songs, and dances were passed on to their descendants through oral traditions. The rich folkloric and metaphorical recollections of the forefathers inspired the descendants to nurture it and pass it on to the upcoming generations.

The Karbi have their own concept of religion; they perform various life preserving rituals at regular intervals. The substructure of the first Karbi village of a 'thousand household' (*don suri*) was described in *Rongkim Alun* (songs about the creation of village) with divine blessings. The verses describe and elaborate the mass relations and chronicle the establishment of the Karbi homeland. During such conversation terms such as 'orality' and 'oral tradition' are used interchangeably. It is observed that right from the cradle to the grave the life of the Karbi is ruled by multitudes of ritual observance which is always accompanied by incantations. Starting from the birth of a child, through the cycle of infancy to the various stages of adulthood until death, the life of a Karbi experiences various 'rite of passage'. Thus, it can be said that the root of the Karbi culture could be detected through folksongs and incantations. The Karbi are 'ancestor worshippers' who perform *'Chojun'*, *'Tirim Kangduk'* (propitiation of ancestors), and *'sar kethok'* (ancestor rice giving ritual), at regular intervals. Such type of rituals of Karbi society requires the name of ancestors of the household to be recited mandatorily in an appropriate sequential order. The rituals are recited by both the male and female genealogies of the household performing the rituals.

Most of the Karbi still practice animism with their cultural and traditional influences. The practitioners of traditional animism believe in reincarnation and honoring the ancestors. The Karbi believe and worship *tirim* (ancestral soul), which continue to play a crucial role in the religious life of the Karbi. The Karbi normally perform rituals to appease the ancestors' souls and supernatural deities. However, the most important rituals known as *tirim kipi* should be done before performing or during the religious practices and communal ceremonies. *Tirim kipi* means appeasing the ancestors' souls. The Karbi have strong beliefs in ancestors' spirits.

Most of the Karbi are very much concerned with their traditional rituals, which they perform before starting any activity, be it work, construction of a house, weaving, or agriculture, where offerings are made to their God or Deities. The Karbi pay obeisance to their forefathers during any ritual as they believe that they must be remembered and that it was through their blessings that they are able to prosper. Furthermore, if they do not remember their forefathers then the rituals will not function smoothly as per their desire. They consider themselves to be the son of '*Hemphu*' and '*Mukrang*'; the literal meaning of '*Hemphu*' is the owner. The Karbi deities are generally divided into different categories. Those categories are as follows.

- Household deities – The deities like *Hem Angtar, Peng*, and so on who are responsible for taking care of the household and protecting the family members.

- Village Deity – This deity is responsible for protection of the village and villagers and keeps away malevolent forces. For example, *Rongker*, *Than Karkli*, and so on.
- Territorial Deities – The deities under this category belong to different territories and are believed to claim ownership of different territories like *Chinthong Arnam*, *Lumbajong Arnam*, and so on.
- Forest Deity – The deity which governs over the forests, hills, and jungles are incorporated under this category, For instance *Habit arnam*, *arlok arnam*, and so on.
- Agricultural deity – This deity is responsible for the good crops and harvests of the agricultural products, for example *Rit arnam*.

Some Cultural Festival and Ritual Practices of the Karbi

It is said that '*Hi-I*' and '*Arnam*' which means 'demon' and 'deity' enjoy an equal status in the Karbi Folk rituals. The presence of deities and their negative 'counterpart' in Karbi rituals reveal their belief in animism and soul (Bey, 2009).

The Karbi celebrate several religious festivals where the rituals are performed for appeasing the deity and the well-being of the community. Some of the religious festivals practiced by the Karbi are discussed next.

Se Karkli

It is a religious ritual of the Karbi; they perform this for the well-being and safe keeping of one's family and field and to protect themselves from sickness and suffering. The Karbi family performs *Peng Karkli* and *Rit Anglong* every year. There are several types of *Se Karkli*, for which the requirement varies; most of the items are collected from the nearby forest, as some plants are considered taboo to grow at home. In *Se Karkli*, the priest called as *Kurusar* resorts to reading the inner organs of sacrificed hens and rooster, goats, or pigs. The organs like gizzard, liver, intestines, and lungs are inspected for reading the outcome of the ritual. Besides reading the entrails, gizzards, livers, and so on, the direction of the wings, head, and position of the sacrificed chicken is examined to read and predict the effectiveness of the ritual.

Tirim Kangduk

The Karbi performs a ritual before any ceremony in remembrance of their deceased family members called *Tirim*. The ritual known as *Tirim Kangduk* is performed to let the ancestors know that they are still remembered. In other words, it means that the ancestors are appeased, and their blessings are sought upon. It is believed that when the *Tirim* are not remembered then the family faces the wrath of their ancestors. For example, rats may chew up their clothes even when kept in a closed box, the granary may be infested with rats, cramps in the hands and legs may occur during a ritual, untimely heavy rainfall and strong winds may blow during any ceremony, and so on.

A banana leaf is sliced, based on the number of family members to be remembered. Certain offerings are made such as *Beng* (usually the tail of a fish, if not available then a dried fish is preferred), strands of meat (pork and chicken), a piece of fish, *Hor-lank* (rice beer), and *Hor-arak* (traditionally brewed liquor), which are placed on all the strands of the banana leaf for the deceased family members.

Chojun

The *Chojun* ritual is also known as '*Arnam kethe karkli*'. It is a ritual which is performed once in three or four years. This ritual is performed for the well-being and safe keeping of the family. Several fowl and a pig are sacrificed during this ritual.

Chomkan

One of the most vibrant festivals of the Karbi, which is a death ritual, is commonly known as *Chomangkan*. The purpose of *Chomangkan* or *Karhi*, a ritualistic and colourful event, is celebrated in memory of the departed soul. It is not only a memorial service for the departed loved ones, but a socio-cultural practice where the entire village and people from surrounding areas participate.

Botor-Kekur

It is a religious celebration which is connected to agriculture. This religious celebration is held for favorable weather and successful crops. The celebration usually takes place in the month of March. It is a communal celebration where the deity of the season/weather is appeased with offerings and chanting.

Hacha kekan

An annual harvest festival, there are two different types of *Hacha Kekan*. The first is observed during the months of December–January while the other is celebrated during the month of September. The *Sokroi Kekan*, which means dancing, is related to the merrymaking and carrying of the harvest, while the other one, called *Hen-up ahi hi ke-en* and usually performed in the month of September, is a ritual where the community gathers bamboo shoots and stores them in a common place – in the big bamboo basket.

Material Cultural Implements Used in Rituals

Bongkrok *(bottle guard)*

Bongkrok is the most important material cultural implement used in rituals among the Karbis. The importance of *Bongkrok* can be found in its necessity in every ritual. The *Bongkrok* is a hollowed up and dried bottle gourd. *Hor lank* (rice beer) is poured inside the *Bongkrok* for offering it to the deity in different rituals. In some rituals *Hor kangthir* (fermented rice cake with water) is stored in the *Bongkrok*. It is used during marriage ceremonies. Every household has their own *Bongkrok*, which are used for different rituals. The *Peng*, who is considered as the household deity and which is found hanging in the entrance of a Karbi household, has a *Bongkrok* on it. Many *Kurusar* always carry their *Bongkrok* in their *Jamborong* (a traditional Karbi bag) wherever they go. There are different types of *Bongkrok*.

Bamboo

Bamboo is found abundantly in all the places of Karbi Anglong, and hence it plays a vital role in the cultural tradition of the Karbi. Bamboo is used as a raw material in almost all the material cultural implements. Some of the material cultural objects made with bamboo and used in different rituals are as follows:

- *A-langpong*: A small bamboo tube filled with water, it is placed in a vertical position near the *Dovan* (mound made with mud which serves as the altar).
- *A-Vophong*: This is the delicate leaves of bamboo which are used in rituals like *Karjong ase*, *vur kematha*, *than karkli*, and so on. (Different types of ritual practices.) The leaves of the bamboo are placed in the *Arnam Avan*, which is a platter with offerings.
- *Anghoi*: The meaning of *Anghoi* is the place to sit (like a stool). A small branch of a bamboo is taken and cut to a small size of approximately 7 cm. The bamboo is split into two halves. It is placed on the *Dovan* before placing the platter of offerings on it.
- *Bi-aput*: This is used as a handle to tie the goat meant for sacrificing. It is taken from a type of bamboo called *Arjang*.
- *Vo-um*: it is a bamboo cage made for storing and carrying fowls for different rituals; the size of the cage varies.
- *Methek*: The branch of a bamboo of around five to six cm is split into half; the split halves are placed at the lower part of the *Dovan*. A burning charcoal is placed on the split bamboo and *Hijung* (a good fragrance resin) is placed on the burning charcoal.

Banana Leaves

Banana leaves have a very significant role in the religious life and material culture of the Karbis. Traditionally during religious festivals or any social gatherings banana leaves were widely used as serving plates and are even used till date. However, its uses are being replaced with the disposable plates available in the markets. The use and importance of banana leaves in different rituals are discussed as follows:

- *Arnam avan*: The end portion of the banana leaf called *Lo alon* is used as the serving platter for the Gods which is placed on the *dovan.*
- *Losen*: A banana leaf is cut into two equal halves and is thrown on the *dovan* before placing the platter for the deities; it is a kind of divination before starting the *Se karkli.*
- *Bong a marthu*: The banana leaf is folded into a conical shape which is then used for plugging the opening of the *Bongkrok*; it works as the cap of the *Bongkrok.*

Sakhi

The *Sakhi* is a small metal ring usually made of brass. It is placed on the middle of *Arnam Anvan* during *Se Karkli.* It is used in some types of divination as well. The *Kurusar* (priest) taps nine times in the *Arnam a-sakhi.* Sometimes the *Kurusar* wears the *Sakhi* on his fingers as a ring.

Him/Sang Aduk

Ths powdered rice is made especially for *Se karkli.* Before the offering and sacrifices are made, the pounded rice powder is mixed with water and the solution is smeared on the *Dovan.* After the sacrifice is done, the remaining pounded rice is then cooked with the sacrificed fowl meat and is served in the feast. This meat cooked with pounded rice powder is called *Sengkunchi Ahan*, which is believed to have a different taste and aroma, which is relished by everyone.

Tari/Noke

This knife is used by the *Kurusar* for sacrificing the fowls during the rituals like *Se karkli*, *Vur Kamatha*, and so on.

Nokjir/Nok

This is a long and heavy-handled sword; it is used for sacrificing goat during different rituals. It is a very sacred sword and is not used at other times. It is also used for divination, which is called *Nok Sanglang.*

Hon Kevai

It is a thread which is tied on the wrist of a person during rituals like *Amenchi Ke Ir.*

Psychological Significance of the Environment in Karbi Society

The intricate web of social connections referred to as a society is achieved through a succession of interactions between individuals to achieve greater means of subsistence and development. Human society and cultural elements are strongly interconnected. The necessities of daily life and means of survival are accumulated in culture. Human existence and social behavior are governed by the norms, values, and conventions associated with culture. Every civilization is subject to an environmental context which serves as the primary source from which the concerned individuals can gather their needs and that creates a bond of dependency. The environment influences society's rituals and customs as well as its economic foundation and material necessities in a variety of ways.

There are many ways to observe how the environment has an impact in tribal societies; their social, economic, and cultural lives. Cultural acts such as songs, dances, proverbs, and riddles reflect their interaction with the environment because they are so closely surrounded by it. Folk cultural traditions serve as a marker of their social harmony and shared values. Folk music both reflects the beauty of nature and expresses a prayer for the general well-being and prosperity of society. Due to their deep ties to nature and the environment, their folk cultural traditions provide more than a means of amusement; they demonstrate a commitment to and submission to nature which influences them psychologically. They pray to nature to guard their assets and their property while continuing to supply them with natural goods. Folk culture encompasses a much larger range of artifacts than just the artistic endeavors and musical performances of the tribal people. Folk-song lyrics refer to the gathering and collection of natural resources to meet the needs of the entire community. The knowledge and customs about how the environment can safeguard the tribal people's existence and how they should use the natural resources in this regard are closely tied to the folk culture's heritage. It prevents them from wasting raw materials and encourages them to resolve environmental degradation. These customs teach one the way of leading life judiciously.

Environmental resources are closely linked to the Karbi's way of life, culture, and economics. Naturally, they desire the natural forces to be merciful to ensure their ability to support themselves. Different natural resources are used as cultural offerings in their rituals. The traditional folk culture demonstrates the enhancement of nature as well as their understanding of the environment and the cultural integration of nature with their way of life. The community's integrity is reflected in their folk festivals, which are a kind of collective devotion carried out by all the community members. Thus, it can be said that the Karbi people's festivals reflect their connection with nature and are an integral element of their cultural cognition. Their aesthetic mind reflects their dedication to the higher power and is closely linked to the natural and supernatural agencies that are thought to govern their sustainability and well-being. Folk, festivals, rituals, the environment, and cultural cognition are all thus closely related, bringing everyone close to home, building an ethos of mutual dependency.

Present Scenario

Cultural ecology points to an intimate relationship between the culture of an ethnic group and its surrounding environment and natural resources. The culture of an ethnic group or community is influenced by his environment and in turn influences the environment. The Karbi is one such community whose life and day-to-day activities are very closely related and attached to the environment. In recent years it can be noticed that large areas of the forest in Karbi Anglong have been cut down due to several factors like rising population, illegal infiltration, developmental projects, and anti-social activities like timber smuggling, illegal coal mining, and so on.

The world now is facing a major problem of environmental destruction; India too is not far behind. Along with it, tribal and rural poverty are an additional burden. The Karbi people of Karbi Anglong have significantly found that the concept of ethno-ecology and indigenous knowledge is deeply correlated with the sustainable utilization of different natural resources. Their traditional occupations have had a deep sense of correlation with the appropriate utilization of natural resources, which were totally organic based, chemical free, and devoid of any hazardous effect of pollution.

Environmental Degradation in Karbi Anglong

It has been found that the forest cover of the district has been declining in recent years. Illegal mining of coal and minerals is one of the main factors along with timber smuggling and deforestation. Diphu, the headquarter of the district, in its journey for expansion and development chopped down numerous trees, which in turn may have led the town to experience flash floods which was not an issue until recently. Due to rapid increase in construction and widening of roads most of the natural drainage systems were destroyed which again is a cause for flash floods.

Cultural Change

According to the Karbi the concept 'forest' meant the resources which were produced in the forest. Forest produced timber and wood. They used to collect timber and wood for house construction and the firewood was utilized for cooking food for the family. They collected bamboo for making a traditional house and it was used in different rituals performed. They collected honey from the forest, and it was stored in the bottle for months and years and they believe that the longer the honey can be stored the better medicinal value will it have. They collected forest resources like edible roots, shoots, leaves, and fruits for self-consumption, ritual purpose, and sometimes to be sold in the local market for sustainability. A variety of bamboo and cane were collected for household-used basketry, weaving loom implement, rituals practice, and agricultural implements and fishing implements. The Karbi collected different parts of plants and herbs from the forest for medicinal use.

With the introduction of cash crops like rubber plantation, betel nut plantation, and others, food crops like rice, sesame, tapioca, maize, yam, and so on, which were grown on shifting cultivation are being replaced. Cultural practices like *Jir Kedam* which is done during the shifting cultivation is slowly diminishing day by day. The Karbi believe in sacred groves too. There are several sacred groves in the district like Ronghang Rongbong, Inglongkiri, Bichikri, and so on. Because of such places the forest and environment of those areas are not disturbed. On the one hand it is keeping the culture and beliefs of the Karbis intact; on the other hand the forest and environment of those areas are also protected from encroachers and other anti-social activities.

Conclusion

Human society is keenly interwoven with cultural aspects. The product of the environment is the economic backbone of society and has a strong relationship with their rituals and customs. The Karbi pay reverence to nature as their protector and the impact of the environment on their social, economic, and cultural life can be noticed from several dimensions. Being closely surrounded in the lap of nature, the Karbi cultural performances like songs, dances, proverbs, and riddles are the reflection of their interaction with the environment. Folk cultural traditions signify their social solidarity and community consensus. The folk tunes reflect the beauty of nature and simultaneously are the prayer for the prosperity and welfare of the entire society. Being closely related with nature and environment, their folk cultural practices are not only for recreation but also for devotion and obeisance to nature.

Folk culture does not imply only the artistic activities and musical performances of the Karbi, rather it bears a much wider aspect. The folk lyrics denote the collection and accumulation of natural products to provide the requirements for the entire community. The folk cultural heritages are keenly related with the knowledge and practices of how the environment can protect the daily livelihood of the Karbi and how they should utilize the natural resources. It prevents them from the wastage of raw materials and implores them to redress environmental degradation.

Among the Karbi people ritual practices are a part of their cultural tradition and social heritage. Particularly regarding the *Chojun* (ritualistic practice), which is performed both at the familial level as well as community level. Irrespective of any gender bias, age, and category, all may participate in it. It symbolizes their collective rhythm of community consensus and social integrity. They show their obeisance and devotion to nature, the souls, spirits, and supernatural forces. This ritual is performed primarily for their social and economic security as well as community sustainability. This celebration is associated with their ethno-cultural identity in a holistic perspective.

At present environmental degradation and sustainable development has become a global emerging issue. In this context the indigenous knowledge can be applied for environmental preservation as well as to protect the environment from its degradation and sustain it for the use of future generations. The Karbi concept of environmental preservation may have some regional variation, but the key concept lies in the issues of safety and security of the environment for all.

It may be summarized that the livelihood of the Karbi people is intimately related to different environmental resources. They have their own concept of ethno-ecology and indigenous knowledge related to the identification and utilization of different natural resources. Such resources related to their rituals bear a special significance to their food and nutrition and cultural traditions. Although the last couple of decades have seen a continuous environmental degradation in Karbi Anglong, threatening livelihood and cultural sustainability, it has been noticed that through folk cultural traditions, Karbi people are trying to preserve their concept of ethno-ecology and are trying to maintain sustainability by going back to their roots.

References

Bey, M. S. (2009). *The Karbi Belief.* Karbi Lammet Amei.
Deogaonkar, S. G., & Deogaonkar, S. S. (2003). *Tribal dance and songs.* Concept Publishing Company.
Herskovitz, M. J. (1955). *Cultural anthropology.* Oxford and IBH Publishing Co.
Tensuba, K. C. (1991). *Genesis of Indian tribes.* Inter India Publications.
Teron, D. S. (2008). *The Karbi Studies Guwahati.* Angik Prakashan.
Vidyarthi, L. P., & Rai, B. K. (1976). *The tribal culture of India.* Concept Publishing.

8 *Luh* and the Monpas of Arunachal Pradesh

Pema Droma and Leeyir Ete

Since the dawn of time, humankind has lived in and around belief systems. These tenets significantly shape the art, literature, and religious makeup of a community. Cultural belief systems not only reflect the traditional and philosophical foundations of a society but are also mediums through which the essence of a tribe's psyche, morality, fears, aspirations, norms and attitude towards life can be captured (Colby, 1966). Mythological belief systems are stories that define one's personal sense of reality and were created to mirror the values of societies in metaphors whilst inculcating an understanding of the unknown in both small and large civilizations. For instance, the population of ancient Greece, a prototype of Republic government, considered thunder to be a bad omen, while the Monpa tribe from Arunachal Pradesh believes that the roaring of thunder is that of a Snake Dragon, considered to be an auspicious entity. The objective of a belief system is to justify the unknown; to explain an enigma.

According to psychoanalysts, folk tales and myths embedded in such belief systems are depictions of psychic repressions of a community's hopes, fantasies and conflicts. As such, the expression of such stories and lores are both projections of the collective unconscious of the human mind as well as a defensive mechanism to release constrained tension (Das, 2014). Furthermore, Jungian interpretations assert that folk tales, legends and myths symbolically create a dialectical and analogical link between the conscious and the unconscious state of mind.

The metaphors in such belief systems not only serve as a source of entertainment, but also provide an individual with an opportunity for moral and personality development by developing one's consciousness. Norton (1999) emphasized this when he asserted that for the healthy social and personal development of an individual, a positive attitude towards one's culture and that of the others is crucial. A psychological understanding of a belief system of a group gives insight into and helps decipher the complexities of human behavior and personality.

Arunachal Pradesh

India's 'Land of Rising Sun', Arunachal Pradesh is the extension of Eastern Himalayas holding its place as the Easternmost state of India. It was born out

DOI: 10.4324/9781003504870-10

of British India's political division NEFA (North East Frontier Agency) on the 20th of February 1987 as a state of India. It shares its international boundaries with China in the North and Bhutan in the West delineated by the McMohan line, and Myanmar in the East. It is the largest state of the seven sisters in North East India with an area of 83,743 square kilometers (32,333 sq mi) and has a population of 1,382,611 with a population density of 17 persons per sq km (Census 2011).

Arunachal Pradesh is a state of geographical diversity, "It possesses snow covered mountain peaks in the Tawang district on one hand, and on the other hand, has foothills and plains in the East Siang and Tirap district on the Eastern side" (Lama, 1999). The state's name was inspired by the historical Indian texts, which name it *Aruna* and emphasize it as the 'land of the rising sun'. The state's diversity is amplified through the hills and the mountain ranges it runs along, that is, the Himalayan ranges (Dafla, Miri and Abor hills), Mishmi hills and the Patkai range. The state accentuates its diversity in flora and fauna with eight wildlife sanctuaries and one reserve forest. The major rivers of the state, namely Subansiri, Kameng, Dibang, Siang, Lohit and Tirap flow through the state whilst breeding the earth with pastoral beauty.

The state is an epitome of ethnic diversity nesting 26 major tribes and 100 sub-tribes, wherein some of the major tribes are Monpa, Galo, Adi, Tagin, Nyishi, Mishmi, Nocte, Wangcho and Apatani and is known to be one of the most multilingual tribal regions of the world. All the major festivals of the tribes are closely associated with the start and the end of farming, these festivals work as celebration of and thanksgiving to nature. "The festivities of these tribes are closely associated with agriculture and thus, a kaleidoscope of cultivation-related dances, songs, rituals, prayers and ceremonies during the local fetes, making it an ideal cultural bonanza" (Bhalerao et al., 2016). Some of the major festivals of the tribes are Si-Donyi, Dree, Solung, Mopin, Nyokum and Lossar.

Most of the tribal societies around the world foster an understanding of man and nature being parts of an inseparable whole, who must live in peaceful co-existence with each other. The existence of similar tales and myths in the belief systems of the Arunachali tribals has attracted the interest of several researchers. The indigenous folk stories of the state – the *Donyi* (Sun), *Polo* (Moon), *Abotani* (first man on Earth) and the animal kingdom – make a good attempt at capturing the essence of social reality, community rituals, morality, norms and the bonding people have between themselves and with nature. However, despite studies on such tribal oral narratives, there still exists a scarcity of investigation from a psychological perspective. As discussed previously, the belief systems of Arunachal Pradesh portray integral information about a community's psyche and their perception of fears, conflicts, desires and virtues.

Monpa Tribe of Arunachal Pradesh

The term Monpa comes from the amalgamation of two Tibetan words '*Mon*' and '*Pa*'. The word *Mon* translates to South and the word *Pa* translates to

People from Tibet; collectively, it translates to 'Lowlanders'. The Monpas are one of the major tribes of Arunachal Pradesh and are one of the most culturally rich tribes of Arunachal Pradesh.

> The Monpas, a Buddhist tribe of the Mahayana sect, are inhabitants of the extreme western-most districts of Tawang, Dirang and Kalaktang. Based on the place of living and the geographical location of these centres they are often called Tawang or Northern, Dirang or Central and Kalkaktang or Southern Monpas.
>
> (Pandey, 1997)

Since Tibet is north of Arunachal Pradesh, the term given to the people rising in the Tawang, Dirang and Kalaktang were aptly named when they were under administrative jurisdictions of Tibet which was taken over by British India with the etching of the McMohan line as a result of the Simla convention.

Although the villages of the Monpa community are considered centuries old, their history is shrouded by the absence of any written documents; "Infromation relating to the origin and migration of the Monks to their present habitat in Arunachal Pradesh is obscure. This is because the Monks were never in the habit of maintaining any written record on their own" (Lama, 1999).

Buddhism to the Monpas is sacred. Their lives are greatly impacted by the principles of Buddhism. Former Lt. Governor of Arunachal Pradesh from 18/01/1979 to 22/07/1981, R.N. Haldipur, IAS, in his article 'Introduction of NEFA' writes

> Among the tribes of this District, Monpas are quiet, gentle, friendly, artistic, and dignified. The influence of Buddhism in their life is marked in every aspect of their lives. A certain gravity and order in their manners, modesty, a consciousness of protocol and the ceremonial daily life impresses one. They follow the Nyingmapa school of Buddhsim and hold in veneration Guru Padma Sambhava. The great Monastery at Tawang is the heart of Monpa life and culture. The sixth Dalai Lama is reported to have come from this region.
>
> (1966)

The festivals celebrated by the Monpas are marked by religious motifs. Lossar, the major festival of the Monpas, dates according to the Tibetan-Buddhist Lunar calendar. Lossar is celebrated for a complete 15 days from the day it starts. It is a festival of harvest and a celebration of the arrival of spring. "The Monpas are very fond of songs and dances. The popular dances are Ajilanum and Sathemine. The Lossar and Torgya are the main festival celebrated as per Buddhist calendar and procedure" (Pandey, 1997). "The Torgya festival is celebrated from the twenty-seventh to the twenty-ninth of the eleventh month. The lamas hold services in the assembly hall on these days on the recitation of several religious texts collectively called Torgya Chakhar" (Sarkar, 1980).

According to Sarkar in 1980, "The Monpas, with the population of 28,209, are the largest Buddhist tribe". As of the 2011 census, the Monpa tribe, with a population of 60,545, is still in majority a tribe with Buddhism as their religion. The history of the Monpa tribe is deeply rooted in the arrival and proliferation of Tibetan Buddhism amongst the tribal folks. One cannot talk about the Monpa tribe without referring to its religious attributions. The reflection of this history in their oral literature and belief system is massive. In art it represents the narrative details in Buddhist aesthetics. Like most societies, the belief system shaped by religion takes a rooted grasp on cultural and societal understanding and the tribe also harbours its own share of myths, folklore and legends like that of the forest, the sacred mountains of '*Ni Dhirshing*' and '*Namchen*', '*Sherlonpoche*' (which means 'the golden elephant'), '*Luh*' (*Nagas*) and animal sacrifices to name a few.

Luh

Luh is a term that defines the creatures of one of the six realms of Tibetan Buddhism. They are believed to be Demigods or Great Snakes (*Nagas*) from the Underworld living parallel to human beings. *Nagas* are believed to be a type of sentient beings belonging to the animal realm and are often depicted in the Wheel of Life paintings based on the Abhidharma literature. These Great Snakes or *Nagas* symbolize the *Luh* in many forms of literature and art. According to the Monpa folk narratives, they can be either vengeful or kind and have the power to harm or bless humans, depending on the conduct of the humans passing them. Their manifestations are believed to take place in the form of trees, animals, river, forest and sometimes even an entire mountain, while in literature and art they are portrayed as mermaids, snakes, and others. The people of the Monpa tribe, to this date, live with this notion of *Luh*, existing in a different dimension.

Luh, in terms of societal understanding are reciprocal beings. This notion instills respect and gratitude within the tribe of the Monpas, but it largely works by creating a psyche of fear in the minds of people who believe in it. It imparts a sense of psychological and emotional understanding of nature. *Luh* is also a term given to a myriad of beings who can be understood as liminal beings who exist as deities, animals, mountains, rivers, forests and even as boulders.

The Abbott of *Thpten Choegyeling Gompa* or Monastery in Kalaktang, in the West Kameng district of the state, Mr. Jingmey Lungtok (aged 53) claims that

> *Luhs* exist where the water originates, especially in the ocean and seas. It is believed that Gombu Lodoe, who is also known as Nagarjuna, brought the Buddhist text 'Boong', from the underworld of Nagas. This text contains twelve books consisting of all the teachings of Buddha. It is read once a year by around thirty monks for the people of Kalaktang.

Mr. Lungtok also claims that the '*Tak-lung Gompa*' (Monastery), a politically and historically eminent edifice situated at a quaint village called Morshing in the West Kameng district, was the residence of the *Luhs*. The shape of the mountain on which the monastery is built is also shaped like the head of the snake and provides a justified imagery to establish a literary memory to the listener. He further states that the legend of the Taklung Gompa speaks of the story of Lama Taipei Dronme (a disciple of the second Dalai Lama, Gedun Gyatso), who tamed a great snake that had harmed several villages and was found drinking water from the river in Morshing. Lama Tanpei Dronme was then believed to have turned the snake into a protector of the Morshing village, and directed it to protect the village and its people. In conclusion he adds, "*When 'Luhs' are prayed to, they reward us with good fortunes.*" The stories and legends of the Taklung Gompa are reportedly very popular and have been heard by almost all the natives of the villages of the district.

The Psychology of Folk Tales

It is of no doubt that the emotion of fear can significantly motivate and direct one's thought, emotions and behavior and, as such, has been continually used as an instrument to control directives within a society. Certain aspects of a society's belief systems contain elements of caution and fear which instill important values of safety and security. Ergo, some cautionary folktales were created to warn people against the different hazards of life whilst also simultaneously controlling undesirable behavior. A similar parallel can be drawn to the stories of *Luh*, wherein the element of fear submerged within the tales has translated into ideologies and actions that promote pro-environment behavior and preservation of nature. A good example of such an inference is an area named '*Wuktsang*' (translated as dark and thick forest) located at the Kharsa village under the Jang circle of Tawang District in Arunachal Pradesh, which is believed to be an abode of the *Luhs*. A native of Jang district of Tawang, Miss Karma Dolma (aged 23), currently a student of Buddhist philosophy, says

> It is believed that Wuktsang is an abode of *Luhs*. It is covered with dark and huge trees and anyone who enters or disturbs this forest is said to have faced bad repercussions like infection on skin or leg cramps.

Due to the fear of the tales, no human being is allowed to enter or litter the forest. They are not even allowed to pick a twig nor pluck a fruit in this forest. This has consequently resulted in the preservation of the natural flora and fauna of the area and kept it safe from the destructive actions of men.

Another instance of the same can be observed in the legends of the Taklung Gompa or Monastery, which has contributed to the practice of snake preservation. Mr. Rinchin Norbu, who has lived most of his life in and around Morshing village (recounting the story his father had told him about the Tak-lung Gompa), said

> It was Lama Tanpei Dronme who turned a huge snake into a protector of the village and on the head of the snake the monastery was built. On a sunny day, one can see a number of snakes coiled around each other. People are not allowed to harm them during this leisure nor do they want to. People are surely afraid of them but mostly see them as the remnant of a being that once existed.

The basis of fear towards nature lies in nature's mysteries. The lack of explanations of occurrences and coincidences involving certain experiences has instilled an emotion of fear towards the *Luh*. A sense of relief and sanity is acquired with the justification of fear-provoking creatures and entities being our protector. Through the act of accepting and rationalizing the paranormal, one can psychologically reassure one's mind about the unknown and perhaps even the dangerous. The following are excerpts from a series of interviews that highlight the fear embedded in the myths of *Luhs*, which has played a significant role in the preservation of forest land by prohibiting human entry.

Mr. Gopal Negi, a Ph.D. student and a translator working in the Central Institute of Himalayan Culture Studies in West Kameng District of Arunachal Pradesh, claims that during his research visits to Tawang he came across Lungla village, an area that has acres of protected and forbidden land and is covered with thick forest and wildlife. He says

> This land was preserved because people are afraid of and respect the deities of the forest. It is said that the forest belongs to the world of *Luh* and should not be littered with remnants of human existence. To this date, it has remained untouched. It is said that many people entered the forest and fell ill and since then not even a twig has been plucked off this forest.

Mr. Sonam Dargey, another native of Tawang, a student of Buddhist philosophy and a teacher at Central Institute of Himalayan Culture Studies, Dahung West Kameng, says

> My friend whose name is Lobsang recently started build a house on a land he owned. Shortly after this he started experiencing severe pain in his leg. He sought several medical help, which did not relieve him off his pain. He then later went to a local priest who said that he had been cursed by the *Luh* because he had been building a house where the *Luh* resides. My friend then conducted a ritual in which he cleaned the entire area and the priest prayed to the *Luh* to forgive my friend and offered milk and flowers to *Luh*. Within two days, my friend was well again.

Mr. Dargey also mentioned that though he had grown up hearing the stories of *Luhs* surrounding human beings, he never believed in them until this event.

Mrs. Dorjee Tsomu of Komlangchen village situated under the Kalaktang circle of West Kameng District says that she had been harmed by the *Luh* in Domkho village when she was a young girl.

> I don't remember what I did exactly to anger the '*Luh*' but I suffered from severe leg pain and the pain only subsided when my father, who was familiar with the ritual, performed it. He made a garland of flower and put it in on a boulder near our house. He cleaned the entire place around the boulder thoroughly and offered milk to the '*Luh*'.

Several such accounts have reportedly occurred which have greatly influenced the way of life in Monpa society. Fear works as the primary emotion whilst encapsulating the belief system of *Luh*; another such important emotion is gratitude, which is believed to have kept the belief system of the *Luh* alive. Mr. Leiki Dondrup (aged 35) a postgraduate in Buddhist philosophy and a native of Dirang village, and Mrs Genden Wangmu (aged 56), a native of Domkho village, both spoke of having heard one such tale during their childhood. The story is of an old lady, who one day found a huge cobra beneath the floor of her traditional hut. She knew that it was the manifestation of the *Luh*. It is said that she immediately cleaned the area where the snake was resting and fed it milk. She then placed a traditional carpet for the snake to rest on. She prayed to the snake to leave the place, urging that it was not clean enough for a deity to live on. The next morning, she checked under her floor and found that the snake had left but had also left behind a heap of gold as a reward for the old lady's kindness.

The old woman in the story is an embodiment and personification of gratitude in the lores of the *Luh*. It compensates for the lack of acknowledgement of a wildlife and its rights. The narrative of this story has travelled to different places and has remained resilient to the passage of time whilst ensuring the proliferation of its teachings to several generations.

This tale of the old lady is an example of the notion that promotes the acknowledgement of the existence of other beings (those who cannot be seen but can be prayed to and granted forgiveness from) within the Monpa culture. This belief system indirectly hints towards an understanding that nature in its entirety does not just belong to the humans but also to other beings. It often generates a psyche of humility and co-existence, resulting in prayers and flower offerings to these entities. The Monpa population is aware of this belief system and to this date continues to practice the rituals highlighted in such tales and stories.

The culture of a community is formed by considering numerous elements of a region like religion, language, climate, geography, castes, sub-tribes, and so on. These factors are intricately woven into and influence the folktales, myths, stories and cultural belief systems that compose a society. Through cultural belief systems and folklore, one can get a glance into the moral values, codes of conduct, traditional practices and dogmas which a society seeks to uphold and safeguard. Consequently, this aids in the preservation of religious, moral, cultural and natural heritage of the society. The folktales and belief systems of the *Luh* are seen to have served this very purpose within the Monpa community. The stories of the Demigod seek to instill a sense of respect, care and gratitude towards nature, which if disobeyed or violated results in adverse and terrifying consequences. The real-life implications of such a practice has truly been constructive for the community, as their ways of natural resource management are vastly reflected in their traditional land and forest management system which include imposition of strict customary laws on the use of medicinal plants, collection of leaf litter for manure and ban on hunting and felling activities on sacred mountains or '*phu*' (Saha & Bisht, 2007).

All elements surrounding the Monpa tribe have an emotional sanctuary attached to it. Irrespective of the reasons for the belief system, the relationship the Monpas share with the creatures living with them in their environment is (what can only be defined as) sublime. Belief systems are foundational in building a community. Its relevance may be questioned in the modern societal architecture, but one has to accept the potential of a belief system to insinuate itself into a culture beyond antiquity. But with the advent of globalization, migration and the penetration of mainland culture into the tribal customs, one can now witness the disappearance and neglect of this ancient belief.

Mr. Sonam Dargey states that "*the belief system of Luh is shrouded with many superstitious elements and though the argument of Luh is strong with the account of co-incidences, there is a population that considers this belief system deeply superstitious*".

Miss Lobsang Wangmu, aged 33, who belongs to the Komalangchen village, while recounting her memories of *Luhs* from her childhood said

> There is a boulder in Bomdila, near our home which was a *Luh*. I remember the area being clean and the boulder being decorated with khattas (local cloth for greeting) but now it has been desecrated with litters and the area no longer looks like what it used to several years ago.

The evaporation of this belief system can be reasoned with the fear of living in the past and to a certain limit, even the greed of wealth. A boulder no longer only represents an idea of *Luh*, but also as a material of economy.

Conclusion

Belief systems and folk tales invoke a group's sense of solidarity and purpose by instilling a shared bond of commonality. This is crucial in strengthening the unity and harmony within the community and can greatly contribute to holistic sustainable development and nation building ideologies by stirring collective social sentiments (Dahal & Bhatta, 2021). Furthermore, it also generates self-awareness by inculcating an understanding of one's social roles and responsibilities and serves as a tool for socialization and connection between generations. The moral lessons enclosed in the cultural beliefs of the *Luh* cannot only serve as an entertaining escape from reality, but it can also be used to promote the same sense of respect, care and gratitude towards nature into the coming future generations of the Monpa tribe to shield the community and the region from the ill-effects of modernization and technology.

Although folk tales and traditional belief systems often represent physically impossible phenomena, their capacity to elicit desirable emotions and impart moral lessons is immense. Folktales, along with being carriers of culture, language, art and literature, address many pertinent social issues and problems. Thus, these should be carefully compiled and introduced into educational settings to convey behavioral expectations or codes of conduct to the younger generation in a more enjoyable way (Dahal & Bhatta, 2021). Perhaps, this could be the ultimate remedial approach to preserving such ethnically rich oral narratives and promoting culturally relevant pro-environment behavior. The eco-centric views and wisdom contained within the folktales of the Monpa tribe can be effectively utilized to promote eco-friendly values and sentiments throughout the state and aid in reviving the dying traditional techniques of natural resource management.

References

Bhalerao, A. K., Kumar, B., Singha, A. K., Jat, P. C., Bordoloi, R., Pasweth, A. M., & Deka, B. C. (2016). *Agricultural folk songs of Arunachal Pradesh*. Indian Council of Agricultural Research.

Colby, B. (1966). Cultural patterns in narrative. *Science*, *151*, 793.

Dahal, H., & Bhatta, B. (2021). Folktales: A moral message from the past to the future generation. *Nepal Journal of Multidisciplinary Research*, 4(1), 31–43. https://doi.org/10.3126/njmr.v4i1.36618

Das, R. (2014). Psychoanalytic study of Folktale. *IOSR Journal of Humanities and Social Science*, *19* (10, Ver–II), 13–18.

Haldipur, R. N. (1966). *Bulletin of Tibetology* (3rd ed., Vol. 3). Namgyal Institute of Tibetology.

Lama, T. (1999). Introduction. *The Monpas of Tawang: A profile*. Introduction, Himalayan Publisher, Arunachal Pradesh. On behalf of Tawang District Youth Welfare.

Norton, D. E. (1999). *Through the eyes of the child: An introduction to children literature*. Merill Publishing Company.

Pandey, B. B. (1997). *Arunachal Pradesh: Village state to statehood*. Himalayan Publishers.

Saha, S., & Bisht, N. (2007). Role of traditional ecological knowledge in natural resource management among Monpas of north-western Arunachal Pradesh. *The Indian Forester*, *133*(2), 155–164.

Sarkar, N. (1980). *Buddhism among the Monpas and Sherdukpens*. Directorate of Research, Govt. of Arunachal Pradesh.

9 Paying Back to Nature: Concerns for Trees in *Hijan Hirao*

M. C. Arunkumar

Man-nature relationship is complex as man, being a biological being, has to adapt to nature and has to depend on nature for their needs. Their life is shaped by the nature in which they live. The much-discussed areas of deforestation, climatic changes, marine changes, rising up of sea level or changes in Himalayan glaciations are in one way or another related with anthropogenic factors.

In his book, *Man and Environment: an introduction to ecology and evolution*, Arthur S Boughey (1971) points out that

> We are the only animal species possessed of a cultural strategy that enables us to escape from the confines of one unique ecological niche, and indeed even from the bounds of all the niches established on this planet. Cultural control of ourselves and our microenvironment permits us to adjust readily to the special requirements of any ecological niche.... Our cultural strategy permits us such rapid adaptation that we can escape the ultimate entrapment that follows commitment to a niche of limited variability, a limitation that has already driven many of our ancestrally related species into ignominious rarity or final extinction.
> (p. 3)

Because of human cultural strategy, man can adapt and survive in different ecological niches of the world.

Man mobilizes resources from nature within cultural parameters as well as protects and preserves the environment so that man can sustain his survival over time. A culture, in the sense of man-nature relation, may be considered either as a mechanism of human adaptation to the existing ecological system or as a means of maintaining an equilibrium between man and nature. In both, man explores nature as well as maintains nature with various cultural institutions, norms and practices. All the cultures have their own strategies to protect and preserve the environment as strategies to their adaptation and survival. The degree of ecological diversity explains the cultural diversity of mankind.

DOI: 10.4324/9781003504870-11

> Cultural systems survive in particular environments because they are adapted to them. In essence, this means that the members of the cultural system are able to survive and reproduce by translating into action the various beliefs and policies contained in the cultural traditions
>
> (Beals, Hoijer, & Beals, 1977)

It is evident that there is a cybernetic relation between man, culture and nature. To maintain the equilibrium of this relationship, man develops certain cultural mechanisms to balance human action and human environment. People employ cultural artifacts – the things they make or shape from natural resources – to adapt and adjust to their environment. Artifacts, however, take on much greater meaning for people than that imparted by their immediate use. Finally, people live in an ecological system related to other organisms within their biosphere. They also possess a cultural ecology or system of knowledge about their relationship with other living forms around them (Spradley & McCurry, 1975: 319–20). With the knowledge of their relationship with nature and other living organisms, man develops a mechanism of controlling their actions and behaviors towards nature and other living forms.

L P Vidyarthi (1963), one of the renowned Indian anthropologists, developed the theoretical frame of man-nature-spirit complex. He explained the mechanism by which forest ecology and cultivating methods shaped the socio-economic life of Maler, a tribe in Bihar. He explained how the spirits are used in the mechanism. His followers, and later researchers who were interested in the relationship between man and nature, studied in different contexts and situations using this concept. P K Singh (1981), using a certain mathematical model, tries to assert that the complex is a constant in all societies living at different technological levels.

Manipur

Manipur, a state in the Indo-Burma region, is home of the Meitei, the dominant group of 36 ethnic groups. The state lies between 23.83 degree and 25.68 degree North latitude and 93.03 degree and 94.78 degree East longitude. Having an area of 22,327 square kilometers, the state has two major physiographic regions: the valley of 1,787 sq km and the surrounding mountainous tracts of 20,089 sq km. As there are mountainous regions all around the valley, almost all the rivers flow down to the valley regions, except a few rivers that flow directly towards Myanmar or Assam. Because of this riverine feature, the valley of Manipur was over flooded most of the time throughout her history. There had been many water bodies in the valley till the recent past.

Because of these topographical features, flora and fauna of the state are numerous. The biodiversity in Manipur in both plain and hill areas is tremendously rich. According to researchers, Manipur's diverse ecosystems include tropical forest, sub-alpine and temperate forests, over 4,000 angiosperm species, 430 medicinal plants, 34 edible fungi species, 500 orchid species,

55 bamboo species, 40 endemic rice cultivars, 160 fish species and 21 migratory aquatic birds among others. "There are still unexplored fauna and flora species in the state," said Senjam Jinus, Assistant Professor of Horticulture and Agriculture Biotechnology, FEEDS Hanbung, Kangpokpi Group of Institute (Shirin, 2021).

The Manipuri civilization has evolved out of the hydrological management of water bodies in Manipur valley and balancing the human exploration of nature and maintenance of the rich biodiversity. Cultivation of various flora and at the same time depending on the biodiversity to meet the needs for sustaining the societies in the state requires a cultural strategy so that the biodiversity never ceases to supply its products for the survival of mankind.

There are a number of ethnic groups in Manipur. The People in the state can be classified into four based on their ecological settings: northern highlanders of higher altitude, southern highlanders of lesser altitude, plain dwellers, and lake-shore dwellers. The population density of the highlanders is very low. They depend either on terrace cultivation or shifting cultivation; their cultivations are rain fed. The ethnic groups in northern highlands traditionally maintain a systematic cultivation in terraces. As villages in the highlands are surrounded by forests and mountainous areas, they depend on forest products.

Moreover, northern highlanders have a peculiar land-use pattern based on their traditions. There are three circles: communal ownership is maintained as forest area belonging to the entire village; clan ownership for agriculture; and horticulture belonging to a particular clan or lineage and individual homestead land. Though there are some variants of this land ownership pattern, the communal ownership of land has a cultural strategy to adapt to the surrounding ecology as described by Pr Dimchuiliu, for Tamenglong District, Manipur. This arrangement is to make a well-balanced relationship between biodiversity preservation and exploration of nature for human life.

In the valley, the people of the two ecological niches – plain dwellers and lake-shore dwellers – are cultivating varieties of crops with wet cultivation. They grow several vegetables as well. The lake dwellers engage in fishing as well as in collecting several aquatic plants for various purposes. They use different types of boats for their activities in the lake. The boats are of different types and size depending on their function. In historical times, the people of plain areas of Manipur used boats for riverine transportation. For making their boats, they had to look for big trees, especially from the nearby hills. They also depend on the nearby forest for wood, cane and bamboo, long grasses which are required for making other cultural items ranging from basketry to houses.

In every Meitei village of plain Manipur, there is a sacred complex of *Umang Lai*, presiding deity of the village. Every year, in a fixed period, the villagers organize *Lai-Haraoba*, the festival of merry making of these *Umang Lais*. As rooted in their belief system, the villages preserve a sacred grove for each *Umang Lai*. *Lai-Haraoba* has different functions in the society, one of which is socialization of the villagers. The festival has a sequence of worship, rituals, songs, dance and recitation of ancient texts. In making the Meitei world-view, *Lai-Haraoba* plays a vital role. Recitation of ancient texts in a sacred complex is the medium of renewing an age-old value system and re-enforcing the cultural themes to the villagers. The themes range from sexuality to social stability to conservation of nature.

Cultural and Psychological Association with Nature

In Manipur, man's knowledge about his relationship with nature and other living forms is reflected in various cultural practices, religious performances and cultural values. Their concern about nature and other life forms is recurrently reminded through various cultural practices and performances. The annual merry-making festivals of their gods and deities in almost all the communities of the state clearly show how much nature is valued in their social and economic life as well as the mechanism of restricting human action towards nature and other living organisms in the given ecological niche.

Hijan Hirao is an ancient text of Manipur. The text was written in verse form. As in ancient times the author and year of writing were not revealed, the dating of the ancient texts is very difficult. Many writers attempted to give a tentative year of the text. Ningthoukhongjam Khelchandra Singh, renowned traditional scholar in the state, in his *Ariba Manipuri Sahityagilitihas*, opines that the ancient text might have been written in between the 5th and 7th centuries, by calculating the year of reign of the king. He calculated the year by cross-referencing many other ancient texts, including the royal chronicle (1992). In a similar fashion, L Ashok Kumar Singh (1992) fixes the year of text writing in the 5th century. However, Kshetrimayum Premchandra Singh (2015) opines that *Hijan Hirao* might have written in the 13th or 14th century by considering the degree of archaism in the language, alien words in the text, subject matter and presence or absence of non-Manipuri phonetic sounds (such as r, b, z, dh).

The rhythm of the text is another quality of *Hijan Hirao*. The rhythm makes the text alive through generations as it is sung on the last day of *Lai Haraoba* (merry making of gods or deities). In every Meitei village, the *Lai Haraoba* is organized annually. The days of it may vary from village to village or from locality to locality; however, it is organized before the agriculture cycle starts and after harvesting. The *Lai Haraoba* is not only of worshiping the gods or deities, but also of socializing the new generation and imparting cultural narratives. In short, the entire knowledge system of the Meitei is recurrently spreading over generations. There are various performances in a *Lai Haraoba* with dances, song and music, rituals and many more activities. Through these performances, the society every year recaptures historically accumulated worldviews and personhood over time. In this sense, the narratives of the *Hijan Hirao* are not frozen in certain ancient texts; the narratives are annually revived so that the ancient values are re-engaged in the body social.

The ancient text was about a boat race, special boat making for the race. The text elaborately describes the process of selection of the tree out of which the boat would be made, the ritual performance for cutting down the tree, the melodrama between nature and the trees. The story starts with –

Torrential rain that poured amidst thunder and lightning
Floods every lane, every meadow of the land.
The Singtha River also is brimmed over
And from its every crevice, every burrow
Gushes down hiding snakes, grasshoppers, and crickets.
Providence brought Luwang Ningthou Punsiba
Towards the furious flow of the Singtha from his stroll
And Lo and behold! A grasshopper crosses the roaring river
Clinging onto a tiny twig, half immersed in the water,
A long-tailed rat performs the same feat
Hanging onto a reed swept down by the Singtha.

(Kshetrimayum, 2015: 35)

Luwang Ningthou Punsiba, after the sight, decided to build a mighty boat. Considering the physical features of ancient Manipur having numerous bodies of water, the need for boats for various practical purposes was pivotal. In Manipur, there were various boats of different shapes and sizes for various human activities in these water bodies. Having an idea of making a big boat, he ordered his artisans to find the biggest and tallest tree for the construction of the boat in mind.

> When the king's men found the tallest tree by the hillside, they performed the customary rituals before felling the tree. As it was believed that the trees and forests have spirits, the woodcutters put their axes on the tree overnight and prayed for permission to cut it

The ritual for felling down the tree is performed –

The earth that bears the majestic tree
Has been cleared and wiped clean by the Luwangs.
The duo, Wangmitkhu and Wangmanao, then calls in the maiba
For the offerings which are to be made
Chosen fruits, and flowers, and rice and eggs,
Pieces of silver and gold
All placed on a plantain leaf
As the maiba chants the prayers.
A fowl thus sacrificed for the regal tree.
Luwangmaiba sings the glory of the tree
And turns to Wangmitkhu and Wangmanao,
"Come and make your mark on this princely tree

For he should know he belongs to you now."
...
Then, the duo turns to the Luwangs and announces,
"The day recedes and the sun prepares for His sleep,
We shall cut the tree when He wakes up again.
Pitch your tents brothers and uncles,
Let us consume the night at the dell near the foothill."
(Kshetrimayum, 2015:37)

The ritual is performed respecting the existence of the big tree as well as for restricting arbitrary human action that may affect the man-nature relationship. Moreover, the text clearly shows that the felling down of plants should not be done in the night time, the assumption of which is that all the living beings sleep in the night; and one should not kill any living beings while sleeping.

The cultural artifacts are made out of human needs. For making these artifacts, of course, man must search for raw materials in nature. However, man must impose himself certain restriction in exploring nature.

In the night before felling the tree, the mother spirit of the forest and the father-rain come to mourn the loss of her child.

In the eerie silence of the night
The tired Luwangs lay sleep
But, Irok Laicham Leima,
Malem Tampak Lairenhanbi,
Mistress of hill ranges and gorges,
Mother of thangthong, samba, feija, feingou,
Takes the handsome and graceful tree
To her bosom in utter pain and agony
And bemoans:
"O my son so grand and gracious,
When you were a sapling
I beseeched you to grow as an ordinary tree;
Misithangyan, thanghon, samba,
Sahikuhi, upan, feija, feingou
Ignoring my advice, you chose to become,
U-ningthou, the King of all trees.
...
And they have made the offerings:
Shiny silver that personifies the father
Red gold from the earth that embodies the mother, and
Placed on Changbi la are eggs on the whitest rice, and
Chosen ripened fruits and the freshly plucked flowers.
O my son, my precious
They have recompensed with gold and silver for your life
They have made axe marks on your body.
(Kshetrimayum, 2015: 37–38)

Again, the father of the ill-fated tree is depicted as the rain. He also was very sad. The mother-nature on the earth and father-rain in the sky were so sad, imagining that their son would be killed next morning. The text describes the state of mind of the father –

His father, the king, who is the rain above,
Whose name derives from the union of
Heavenly showers and earthly soil
That gives birth to serene springs and beautiful brooks
Thus, known to all as Irok Ningthou
Who resides with grace at the high ranges
Cries in lament as the thought of his son
About to be hacked at daybreak.
......
O magnificent son, pride of any father,
How do I endure this grief, this void?
I am undone."
...
Filled with benevolence for the dearest son
The grieving parents have their last look at him
And return back as the Sun dawns upon earth.

(Kshetrimayum, 2015: 40)

The latter part of the text deals with how the tree was cut down, the big boat built and how the boat was used. Meaning the boat was built out of the fallen tree, with various rituals and human activities and later became a cultural artifact. A natural object is turned into a cultural object. In this transformation, we see a chain of human need, desire, natural object, respect to the natural object and the becoming of the object into a cultural object. In this change, we observe how man respects nature for which he pays back the value of the object lost by nature.

Here two things are obvious to show that there were cultural practices to pay back the value of the tree before felling it down. The king's royal priest performed certain rituals and prays to nature to give the 'giant tree' and paid the 'gold and silver' as compensation to nature. Nature was represented by mother queen and mountain king. Though the preceding translation was done in a simplified manner for making the reader understand easily, the piece of translation from an archaic Manipuri does not fail to show the relationship between mother-nature and her children. Here is the balance between man's exploration of nature for his cultural artifacts as well as his imagination of the man-nature relationship and relationship between man and other living organisms. Mother-nature is personification of the spirit that controls human action to destroy the trees (and also other organisms) at random.

The text, *Hijan Hirao*, of course, has many dimensions – from ideology to state administration of religious practices. Besides the political and social ideology embodied in the text, here one is trying to locate nature in human adaptation to their ecological niche. As man has to explore nature in search of natural objects for making their cultural artifacts, man has to conserve nature. Again, towards the end of the text, it is said that "*We are not rowing the boat, we are rowing the Divine King and Divine Queen.*" The boat is once again mythicized so that the cultural artifacts are transformed to objects of spirits. These mythicized elements are part of man's larger strategy of human adaptation and sustainable livelihood. This ingrained psychological mechanism manifested as a cultural norm of conservation of ecological niche and the biodiversity therein is internalized recurrently, mainly by the annual festival of *Lai Haraoba*. This is the complete cycle of man-nature-spirit complex.

On the last day of every *Haraoba*, there is a series of rituals while reciting the *Hijan Hirao*. These rituals are performed by *Maibas*, the priest. Because of this process of socialization, nature is considered as an entity to be conserved for human existence. Hence, one should not cut down any plant or kill any animal in the night. If one has to pluck leaves or flowers, he should clap three times and say, "Wake up" three times. If one has to cut down a big tree, he should perform a ritual and make some offerings to pay back for the tree. Many taboos are imposed on human behavior in relation to extracting resources from nature and other living organisms through these rituals and imparting cultural values to the natural objects.

The understanding of the rituals behind the festivities may not lie solely in the practices. The relationships depicted, the emotions shared, the behavioral modifications all indicate that the bond that man and environment share is entwined in the way of one's life. These rituals that have been told and re-told in the form of stories act as a form of transfer of learning through generations. The purpose may be to create awareness about the depth of the relationship that was, is and will be shared between man and his environment.

Conclusion

Hijan Hirao, the text, is performed in the *Lai-Haraoba* every year to socialize man or to remind the people the main theme and sub-themes of the text. Besides social and political themes, one of the sub-themes is the conservation of the environment. The man, while extracting natural resources, must repay the debt to nature; ritually the value of the trees and other living beings in nature is recurrently taught. Man, as being the only conscious animal in the world, must understand the pain and agony of the flora and fauna in nature. The ancient text, in its sub-text of conservation of natural resources, restricts human action on the environment. The repaying of the values of natural objects to nature and ritual restriction on human action in extracting natural resources are internalized in the human action and behavior through these ritual processes. Unlike the positive cultural practices of preserving sacred groves around the shrines or temples of deities, the performance of *Hijan Hirao* in the annual merry making of the deities or gods in these groves aims to create environment or nature as part of cultural values, and hence part of intangible heritage regarding conservation of environment. In Manipur, there are many sacred groves, preserved by the people. The area of sacred groves ranges from a few square meters to several hectares. They are situated in different altitudinal gradients and form 1,542 natural ecosystems which help in conservation of biodiversity (Khumbongmayum et al., 2005). A variety of cultural, religious and traditional beliefs and rites of ancient Manipur are associated with their love and care of wildlife contributing to their conservation (2005).

Because in Meitei society *Lai-Haraoba* is performed every year in every village in and around the sacred complex of the deity, people have a positive practice of preserving sacred groves; these groves are associated with their faith in the deities. On the other hand, the internalization process of the *Hijan Hirao* theme takes place when it is performed in the same *Lai-Haraoba* year after year. Therefore, preservation of sacred groves and internalization of the theme of conserving the environment are but two sides of same coin. The sub-theme of conservation of environment in the *Hijan Hirao* is thus related with the positive practices. While discussing the degradation of sacred groves, Khumbongmayum et al. (2005) expresses that it is sufficiently clear that myths and beliefs associated with the sacred groves, which used to be followed strictly in earlier days, have been eroded during the last few decades and the groves no longer enjoy the same privilege as they did earlier. The linkage between the preservation of sacred groves with the stories in the ancient texts is very obvious. The driving force of the positive action of preserving these sacred groves was the ancient texts such as *Hijan Hirao* or the cultural themes which were reflected in various performances, including those of *Lai-Haraoba*. Unfortunately, due to modernization and other social factors, the age-old themes and practices are degraded or less valued these days, hence the report of the degradation of the sacred groves.

The large-scale degradation of the environment, may it be caused by shifting cultivation of highlanders, deforestation because of commercial ventures of some individuals or of developmental works, road construction or large scale encroachment of water bodies, is a sign of imbalance between human needs and natural resources. It is a sign of imbalance in the man-nature-spirit complex. It is directly or indirectly linked with man's loss of their once-zealously guarded values of nature. As the organic link of these values to present day social life is disturbed by the over-demanding human society, the cultural values of nature cannot be manifested in the positive efforts of preserving nature. It must be understood that the mechanical efforts of preserving nature or sacred groves cannot be sustained as the driving force of culture has lost its authority. Therefore, the ancient texts and cultural performances in which these texts are expressed should be re-evaluated in a larger frame of conservation and preservation of environment. Even environmental education requires these age-old texts in which the present-day message is embodied. The *Hijan Hirao* may be a solution if the value of paying back to nature, which is at random disturbed by several human demands, is reinforced.

References

Boughey, A. S. (1971). *Man and the environment: An introduction to ecology and evolution*. Macmillan Publishing Co.

Beals, R. L., Hoijer, H., & Beals, A. R. (1977). *An introduction to anthropology* (5th edn.). Collier MacMillan.

Khumbongmayum, A. D., Khan, M., & Tripathi, R. (2005). Sacred groves of Manipur, northeast India: biodiversity value, status and strategies for their conservation. *Biodiversity and Conservation*, *14*, 1541–1582. https://doi.org/10.1007/s10531-004-0530-5

Kshetrimayum, P. S. (2015). *Hijan Hirao: Text, context, and translation*. Sahitya Seva Samiti.

Ashok Kumar, L. (1992). *Ariba Manipuri Sahitya Paring: 18 suba chahichaphao-bagimamang*. Shrimati Yumnam Suchitra Devi.

Khelchandra Singh, N. (1992). *Ariba Manipuri Sahityagi Itihas* (3rd edn). Divine Book Mart.

Shirin, B. (2021). *Manipur's biodiversity facing degeneration*. Imphal Free Press, 22 Jan 2021. https://www.ifp.co.in/1473/manipurs-biodiversity-facing-degeneration. Retrieved on 9 December 2022.

Singh, P. K. (1981). Nature-man-spirit complex: A constant? In R. S. Mann (Ed.), *Nature-man-spirit complex in Tribal India*. Concept Publishing Company.

Spradley, J. P., & McCurry, D. W. (1975). *Anthropology: The cultural perspective*. John Wiley & Sons.

Vidyarthi, L. P. (1963). *The Maler: A study in nature-spirit-man complex of a Hill Tribe in Bihar*. Bookland Private Limited.

10 The Sacred Groves of Meghalaya

Baharul Islam Laskar

Indigenous knowledge systems (IKS) are always found in all societies. However, they vary widely across time and space. IKS comprise knowledge developed within indigenous societies, independent of, and prior to, the advent of the modern scientific knowledge system. IKS refers to the understandings, skills, and philosophies developed by societies with long histories of interaction with their natural surroundings. This knowledge is integral to a cultural composite that also comprehends language, systems of classification, resource use practices, social interactions, rituals, and spirituality. These are important facets of the world's cultural diversity and provide a foundation for locally appropriate sustainable development. The United Nations Educational, Scientific and Cultural Organization (UNESCO) further emphasises that the local and indigenous knowledge systems attempt to reinforce indigenous people and local communities, foster transdisciplinary activities with scientists and policymakers, and adopt methodologies to add emphasis on climate change effects, adaptation, and mitigation.

Sacred Groves

Sacred groves are traditionally managed community forests all over the world, where indigenous communities live for centuries. These are tracts of virgin forests that have cultural or spiritual significance for the people who live around them. They have been protected by communities around the world for various reasons, including religious practices, burial grounds, water shed value and so on. Sacred natural sites such as forest groves, mountains, and rivers are the world's oldest form of protected areas and play a vital role in the informal conservation of both biological and cultural diversity (Gu & Chen, 2011).

Many societies around the world, due to their spiritual beliefs, have granted the special status of sacred places to natural areas like mountains, rivers, lakes, caverns, forest groves, coastal waters, and even islands (Oviedo et al., 2005). Their holiness is attributed to a variety of factors, including the presence of deities and ancestor spirits, the availability of healing water and plants, points of touch with spiritual truth, locations of revelation and

DOI: 10.4324/9781003504870-12

transformation, and so on. Further, many of the sacred natural sites have historical significance and are not static in time or space, that is, new sites can be identified in response to changing scenarios. Brahma et al. (2014) considered that religious and traditional rituals related to gods and goddesses performed in various ways replicate the connection of human and species conservation since it serves as a means for conserving biodiversity intertwined with religious practices.

Sacred groves are considered a rich and dynamic source of ecological, holistic treatments for human ailments. They serve as pools or servers for extinct species of plants, restoring some of the rare plants and may provide emergency food during periods of drought, crop failure, and food crisis. The ecological relevance of sacred groves includes biodiversity conservation, rainwater harvesting and aquifer recharge, and soil conservation. These natural advantages contribute significantly to flora and fauna variety.

Sacred groves are an integral part of culture and tradition. These groves are managed by ethnic groups of people and in most cases have some strict customary laws which prohibit any form of harm to the groves. It has great ecological significance, which is reflected in the protection and conservation of the sacred groves, notably having nutritional and medicinal importance (Jagdale, 2021; Khan, 2022). According to Pooja Sah (2022), the interactions between physical and biological elements of the sacred groves ecosystem support: (a) moderation of the climate region's regulation of local hydrology, (b) maintenance of nutrient cycles and soil fertility, and (c) delivery of aesthetic beauty to the landscape. She comments that the safeguarding of sacred groves is part of tribal religious and social ethos, as well as culture, yet rapid changes provide significant obstacles to conserving such ecological hotspots.

The perceptions towards environment and development have changed over time and it has been well recognised that development is the process of furthering people's well-being. Good development involves increasing the asset base and its productivity; empowering poor people and marginalised communities; reducing and managing risks; and taking a long-term perspective regarding intra- and inter-generational equity (Bass, 2006). Thus, the interdependence and relationships of people, vis-a-vis sacred groves or the environment, are needed to be better understood and the present threats need to be addressed. As the environment affects the actual basis of human well-being, it is a concern of justice to deliberate the impacts of environmental degradation and attempt to minimise harm for both present and future generations. In this context, it has been argued that a "global ethic" is required to address the glitches of the 21st century (Singer, 2002).

Meghalaya

Meghalaya was created as a state on 21 January 1972, before which it was a part of Assam. The name Meghalaya means "*the abode of clouds*". This state is bounded to the south by the districts of greater Mymensingh and the

Division of Sylhet, in the west by the Division of Rangpur of the People's Republic of Bangladesh, and to the north and the east by India's Assam state. The capital is Shillong, known as the "Scotland of the East". Meghalaya has a total population of 2,966,889, of which 79.9% are residing in rural areas and 20.1% in urban areas (Population census, 2011). English is the official language of Meghalaya. About 70% of the state is forested, which is notable for its biodiversity of mammals, birds, and plants. It is basically an agricultural state with about 80% of its total population depending on agriculture for its livelihood (Laskar, 2021).

Meghalaya is the native land mainly of the Khasis, the Jaintias, and the Garos. The Garos inhabit Western Meghalaya, the Khasis in central Meghalaya, and the Jaintias in Eastern Meghalaya. The Khasis, the Jaintias, and the Garos have matrilineal families and societies, where the descent is traced through the mother. The youngest daughter inherits a major share of wealth and takes care of her parents, though there are changes in such practices. Other communities of the state include the Assamese, Bengali, Koch, Biates, Rajbongshi, Boro, Hajong, Dimasa, Hmar, Paite, Kuki, Lakhar, Karbi, Rabha, Nepali, and others. There are various festivals celebrated by the people of Meghalaya. Some of the important festivals are Bam Khana, Shad Suk, Shad Nongkrem, Behdienkhlam, Seng Kut Snem, Wangala, Christmas, and others (Laskar, 2021).

Sacred Grove of Meghalaya

In Meghalaya, there are legal safeguards for the sacred groves. These are declared under the United Khasi and Jaintia Hills Autonomous District (Management and Control of Forests) Act, 1958, and the Garo Hills Autonomous District (Management and Control of Forests) Act 1961. According to these acts, the sacred groves are to be managed and controlled by the *Lyngdoh* (priest or religious head) or other person(s), who are delegated with the religious ceremonies for the locality or village, in harmony with the customary practices and rules outlined by the Executive Committee of the respective Autonomous District Council (MBB & WII, 2017).

Religious and traditional beliefs, cultural mores, and practices play a crucial role in the conservation of the environment and biodiversity. Sacred groves and indigenous and community-conserved areas are repositories of biological and cultural diversity and represent communities' religious values, customary rules, institutional fabric, traditional knowledge, and conservation culture (Arjjumend et al., 2017). These communities practice conservation for a variety of economic, social, cultural, spiritual, and aesthetic reasons. There are customary practices and laws across regions and communities associated with the sacred groves that have implications for the conservation of the environment. Traditionally the tribal communities in Meghalaya have been preserving small patches of virgin forest since time immemorial based on their religious belief. They declare a patch of forest near the villages as sacred and protect it on the grounds of religious and cultural beliefs. These sacred groves are owned by individuals, clans, or communities, and are under the direct control of the clan councils or local village Dorbars/Syiemships/Dolloiships/Nokmaships (Kharkongor & Tiwari, 2016).

They believe that the sacred groves are the abode of deities like *Ryngkew*, *Basa*, and others, that confer well-being to the people, their cattle, and land and keep the evil spirit away. People protect sacred groves to avoid the perceived anger of the resident God, in which all forms of vegetation belong to the deity. They believe that the Sylvan deities would be offended if trees are cut and twigs, flowers, fruits, and so on are plucked. Religious rituals like *Ka leh-niampirda* are performed to keep the deities happy. These groves are considered as one of the most species-rich areas and the mythological stories and indigenous knowledge associated with the groves are the principal factors in preserving the sacred groves (Jeeva et al., 2006). Therefore, their religious beliefs and taboos are contributing to the protection of the sacred groves.

A review of available data shows sacred groves are observed in five of the districts in Meghalaya, that is, West Khasi Hills, Ri Bhoi, East Khasi Hills, Jaintia Hills, and West Garo Hills.

Magnitude and Distribution

Sacred groves were reported from many countries including Asia, Africa, Europe, and the Americas but at present these are mostly found in Africa and Asia. A review of the literature discloses that sacred groves are a global phenomenon with a high resemblance in the cultural and ecological extents of their management in Africa, Asia, Europe, Latin America, and Australia (Vincent et al., 2019). In India, these are reported from the North East Himalayan region, Western Ghats, Eastern Ghats, Coastal region, Central Indian Plateau, and Western desert (Ray et al., 2022). These groves are scattered all over the country and are protected by the local people, which is an age-old practice among many tribal communities. In India, estimates of the number of sacred groves are quite different. For example, according to one estimate, there are more than 13,000 sacred groves (https://www.aerfindia.org/grove.html), while another estimate puts India with over 100,000 sacred forests (Ormsby & Bhagwat, 2010).

A field-based study in Tamil Nadu analyses the religious and social context in which sacred groves make meaning for the villagers who maintain them, and explains the fascination for Indian environmentalists, who see the sacred groves as an exemplary form of traditional community resource management (Kent, 2013). The age-old system of every village having a temple, a tank, and related sacred grove explains the ancient method of water harvesting and distribution and may be considered as the backbone of the village economy, where people are prohibited from felling trees which is considered as taboo (http://old.forest.kerala.gov.in.).

North East India (NEI) is known for its sacred forests, called by different names in different states. These forests are protected vegetation and are an integral part of the social system. Each forest system is unique regarding cultural and ecological aspects. Data on sacred groves in NEI reflect that there are 865 such forests covering an estimated area of 27,082.42 hectares constituting 70% of the total area covered under sacred forests in India (https://nenow.in/north-east-news/sacred-forest-its-role-on-environment.html). Data further shows that there are around 125 sacred groves in Meghalaya, with varying areas (http://www.megforest.gov.in/forest_sacredgroves.html). A few selected sacred groves with an area of 100 and more hectares in Meghalaya is shown in Table 10.1. There are 23 such big sacred groves in the state, of which the majority of 12 are in East Khasi Hills District alone, while there are eight such sacred groves in West Khasi Hills District, two in Jaintia Hills, and one in Ri-Bhoi. The *Maw Kyrngah* sacred grove in West Khasi Hills has the highest area of 1200 hectares, followed by *Pahamodem* sacred grove in Ri Bhoi (900 hectares), and *Raid Shabong Law Adong* sacred grove in East Khasi Hills (700 hectares). However, there are many small sacred groves (less than 100 hectares) in most of the districts.

Table 10.1 Selected sacred groves (100 and more hectares) in Meghalaya

Sl No	*Name of Sacred Grove*	*Location*	*Area*
1.	Maw Kyrngah	Wmwai	1200
2.	Pahamodem	UmsawNongkhrai	900
3.	Raid Shabong Law Adong	Pynursla	700
4.	Diengkain	Wmwai	400
5.	Law Adong	Mawsmai	400
6.	Umtong	Wmwai	400
7.	Law Kyntang	Shanpung	400
8.	Law Lyngdoh, Rangmaw	Maharam Syihip	400
9.	Law Kyntang, Mawlangwir	Maharam Syihip	300
10.	Law Dymmiew	Sohrarim	200
11.	Law Adong Lyngdoh, Mawlong	NongkhlawSyiemskip	200
12.	Law Lyngdoh, Kinglang	Maharam Syihip	200
13.	Kharai Law Lyngdoh	Nongkhieng	150
14.	Kynsang	Mawlong	150
15.	KhlooPaiu Ram Pyrthai	Jowai	150
16.	Law BleiBeh	Mawsmai	120
17.	MawlongSyiem	Mawsmai	120
18.	Law Lyngdoh Lyting Lyngdoh	Lyntilew	100
19.	Law Lyngdoh Mawshun	Mawshun	100
20.	Umkatait	Dieng Ksiar	100
21.	Law Kyntang, Mawten	Maharam Syihip	100
22.	Law Kyntang, Whawiaw	Maharam Syihip	100
23.	Nonsynrih Sacred Grove	Nonsynrih	100

Sacred groves are classified in different places and among communities on different bases. In Khasi Hills the sacred groves can be classified into three categories (i) *Law Lyngdoh* are the groves ruled by the priest, *Lyngdoh*; (ii) *Law Niam* are the places of traditional religion, *Niamtrai*; and (iii) *Law Kyntang* are the groves where the area is under control of the village headman (MBB & WII, 2017).

In Jaintia Hills, sacred groves are known as *Khlaw U Blei* or *Khloo Blai*, and in Garo Hills are known as *Asong Khosi*. The stringent customary laws associated with the protection of sacred groves have resulted in better conservation of a wide variety of flora and fauna, and these groves play a significant role in environmental enrichment as many of these sites form vital watersheds and catchment of many rivers and streams (MBB & WII, 2017).

Threats and Challenges

The combination of nature, community, the transfer of traditional knowledge, and the sacredness of the grove, is accountable and signifies the purity of the sacred sites. Natural and social scientists, humanities professionals, and anthropologists are exploring the networks among sacred places, communities, cultures, religions, biodiversity, conservation, and government (Guha, 2000; Sarkar, 2005; Mishra et al., 2004; McLeod, 2002; Mallarach, 2007; Malhotra, 1998). Coggins and Chen (2022) examine the sacred forests of Asia and engage with lively newfangled intellectual discourses on the nature of sacred space, place, landscape, and ecology in the context of the sharply contested ideas of the Anthropocene. They deliberate on the diversity of connected cosmologies, ecologies, traditional local resource management practices, and environmental governance systems developed during the pre colonial, colonial, and postcolonial periods. Chanda and Ramachandra (2019) exert that sacred groves in different parts of India have diverse cultures, religions, rituals, festivals, and traditions associated with the people living in these areas.

Most of the tribes in Arunachal Pradesh have been protecting and conserving sacred groves based on their religious culture, beliefs, totem, taboos, and festivals (Chaudhry & Murtem, 2015). A few of the sacred groves are attached to the Buddhist monasteries, called Gompa Forest Areas, and these sacred groves are dedicated to local deities (http://cpreecenvis.nic.in/Database/Arunachal_Pradesh_878.aspx).

In Assam, forest-dwelling tribes such as Bodo, Rabha, and Dimasa, inhabiting the plains and hills, are maintaining sacred groves (http://cpreecenvis.nic.in/Database/Groves_811.aspx). Vibha Arora (2006) fosters the socio-ecological debate on conservation, socio-religious fencing, and the mediating role of the state and opines that sacred groves and landscapes are often perceived as a case of indigenous forest management practices and the antithesis of the sanctuary rationally managed by the forest department of the government.

These groves are considered among the most species-rich areas for plants, birds, and mammals; and unlike in most parts of India, most of the NEI forests are owned by communities and individuals, and not by the government (https://megbiodiversity.nic.in/sacred-groves). In Meghalaya, like in other places, people believe that the sacred groves are the abode of deities, which grant the welfare of people, cattle and land, and the evil spirits are kept away. People believe that the deities would be insulted if trees are cut and twigs, flowers, fruits, and so on are plucked. These sacred groves are one of the greatest species-rich areas, and the indigenous knowledge system linked with the groves is the primary cause influencing the preservation of these groves (Jeeva et al., 2006). The tribal communities of Meghalaya have a tradition of environmental conservation based on various religious beliefs (Asokan et al., 2015).

There are many causes for the degradation of sacred groves, which include people's changing attitudes, erosion of traditional beliefs, and human impact over the years. Singh et al. (2017) opine that their conservation would not be possible without the active participation of the local people, which should include improving living standards and providing aids of conservation to them. Oviedo et al. (2005) state that sacred natural sites are integral parts of ethnic identity and play a key role in traditional cultures and lifestyles and used to be controlled by the community. They also opine that as people are becoming gradually vulnerable to political and economic pressures, without the security of tenure and active participation in decisions that affect them, it is impossible for such communities to effectively protect their lands and resources. Sharma and Devi (2014) highlight various threats faced by the sacred groves like construction activities, grazing of live stocks, and modernisation.

In a study conducted in the West Khasi Hills and East Khasi Hills districts of Meghalaya, it was observed that many sacred groves have been enormously besmirched, and a considerable number have lost their sacredness with time. Some of them have been transformed into community forests while others have been commercially exploited. Further, the beliefs and traditional institutional mechanisms relating to the conservation of sacred groves are corroding and the science behind the concept is not percolating among the communities who had maintained the sacred grove since time immemorial (Kharkongor & Tiwari, 2016).

In another study, strategies identified for effective management of sacred groves for the conservation of socio-cultural and biological diversity existing in them include

- A self-imposed complete ban on the removal of biomass for maintaining the sustainability of the ecosystem in perpetuity,
- Creation of awareness among local people and all stakeholder groups,
- Identification of the type of contribution a stakeholder group can offer in managing the sacred grove,
- Encouragement of all stakeholders in the management of sacred groves taking into consideration the wisdom and interest of the major stakeholder groups (Chandrashekara & Sankar, 1998).

The challenges of preservation and promotion of sacred groves in the context of NEI and particularly in Meghalaya are somewhat different. The forest cover in all the states of the country is decreasing gradually, due to many factors. However, the situation in NE India, particularly in Meghalaya, is better. The forest and tree cover in the state is 79.37% covering 17,803^{2} km, making it 2.26% of India's forest and tree cover. The state ranks in the 4th position in terms of the percentage of forest cover in the country. However, due to the prevailing land tenure system in the state, only 1,145.19^{2} km of forest areas (5.10 % of geographical area) comes directly under the control of the state Forest Department and the rest of the forest areas belong to communities, clan, and private people and District Councils (https://megforest.gov.in). So, there is an urgent need for the preservation and conservation of sacred groves in the state for the interest of the people, state, and nation.

Suggestions for the Conservation of the Environment

The sacred groves and the IKS attached to them are contributing immensely to preserving and protecting the environment. At the same time, people are attached to it through cultural and livelihood interdependence. However, due to the impacts of modernization and other related factors, the sacred groves and the concerned people are facing many threats and challenges. Some of the threats to sacred groves identified in the context of Meghalaya (https://megforest.gov.in/forest_sacredgroves.html) are as follows:

- Changes in the tribal way of life with the erosion of traditional values;
- The area under sacred groves is shrinking and quite a few have been turned into degraded forests; and
- Some sacred groves have fallen prey to encroachments and degradation.

The modern way of life has contributed to more individualism and materialism, erosion of cultural and community values, migration to urban areas, less importance on traditional livelihoods, and so on. It is thus pivotal to do ethnographic and ethnomethodological as well as survey research on the issue to protect and preserve the natural interdependency of the people and the sacred groves in the ecosystem. In this context, both civil society and the government need to play active roles.

It may be noted here that the world economy has been featured by growing globalisation, which is urging the growing integration of the global economy through trade and financial flows, and in the integration of knowledge through the transfer of information, culture, and technology. Governance has become globalised, with increasingly multifaceted inter-state interactions, and with a growing role for non-state actors. The sacred groves and the livelihoods of the local communities can be promoted using the fruits of globalisation. In such sacred groves, varieties of flora and fauna are available which can be used for medicinal and other marketable purposes, not affecting its biodiversity. It is suggested that emerging communication and surveillance systems could be employed to acquire information or data on a real-time basis and be able to disseminate information as and when it is required. Application of mobile or wireless computing could be pursued for on-spot field identification and data comparison. The development of such a factual database would support the development of strategies for the conservation and protection of these unique heritage ecosystems (Gaikwad et al., 2004).

Further, it is observed that community-based programmes can be markedly operative in handling natural resources, providing basic infrastructure, or ensuring primary social services. ESSDN (1995) suggests that a successful strategy of community-based development requires tapping into local needs, understanding, and building on the strengths of existing institutions and defining the changes needed in intermediary implementing agencies to support community action. Therefore, the local communities should be taken into utmost consideration to protect and preserve the sacred groves, for the interest of all. For example, among the Monpas of Tawang in Arunachal Pradesh, if a grown-up tree is cut from a sacred grove, the person concerned is required to plant ten saplings.

Conclusion

The interplay of cultural practices and conservation of the environment forms the backbone of societies that recognise the value of the ecological balance. These practices shape the biodiversity and landscapes which maintain variations of the ecosystems and sustain local livelihoods. Behaviour of an individual reflects one's cultural values. Culture reflects how humans think and their perspectives towards events or situations. In the context of the collectivist communities of Meghalaya, sacred groves were maintained because of the people's cultural and spiritual beliefs, which ensured that the environment is closely entwined with the cultural beliefs, taboos, and practices. This is in conformity with research by Komatsu et al. (2019), who examined whether individualistic countries would have a higher environmental impact compared to collectivistic countries. Their findings indicate that people living in collectivist countries tend to have a higher-level responsibility for environmental problems.

Maladaptive human behaviour is the cause of most environmental problems (Maloney & Ward, 1973). This idea was supported by Oskamp (2000), who claimed that the major culprit for environmental problem is human behaviour. Cultural sustainability is all about ensuring that efforts are made for the preservation of cultural elements as cultural practices are innately related to environmental sustainability and that inevitably ensures social and economic sustainability.

References

Arjjumend, H., Konstantia, K., Fagan, H. G., & Shibata, S. (2017). International overview of sacred natural sites and indigenous and community conserved areas (ICCAs) and the need for their recognition. *Global Environmental Studies*, *13*, 33–56.

Arora, V. (2006). The forest of symbols embodied in the Tholung sacred landscape of North Sikkim, India. *Conservation and Society*, *4*(1), 55–83.

Asokan, A., Chouhan, S., & Singh, V. (2015). Sacred grove—A nature's gift—as a Remedy for human ailments, a biodiversity reservoir for restoring indigenous traits for endangered listed plants—A review. *Scientific Research*, *2*(7), 1–13. https://doi.org/10.4236/oalib.1101517

Bass, S. (2006). Making poverty reduction irreversible: Development implications of the millennium ecosystem assessment. In *IIED environment for the MDGs briefing paper*. International Institute on Environment and Development. https://www.iied.org/11050iied

Brahma, J., Singh, B., & Mandal, T. (2014). Sacred groves: An analysis made in the cultural perspective within BTC, Assam, India. *Journal of Biological & Scientific Opinion*, *2*(5), 320–323.

Chanda, S., & Ramachandra, T. V. (2019). Vegetation in the sacred groves across India: A review. *Research & Reviews: Journal of Ecology*, *8*(1), 29–38.

Chandrashekara, U. M., & Sankar, S. (1998). *Ecological and social importance of conservation of sacred groves in Kerala*. Kerala Forest Research Institute.

Chaudhry, P., & Murtem, G. (2015). Role of sacred groves, value education and spirituality in conserving biodiversity with special reference to Arunachal Pradesh state of India. *International Journal of Society Systems Science*, *7*(2), 151.

Coggins, C., & Chen, B. (Ed.) (2022). *Sacred forests of Asia: Spiritual ecology and the politics of nature conservation*. Routledge.

ESSDN (1995). Social Development Notes. 27361, Note No. 4. Environmentally and Socially Sustainable Development Network.

Gaikwad, S. S., Paralikar, S. N., Chavan, V., & Krishnan, S. (2004). Digitizing indian sacred groves – An information model for web interfaced multimedia database. In Ghate, V., Sane, H., & Ranade, S. S. (Eds.), *Focus on sacred groves and ethnobotany* (pp. 123–128). Prisam Publications.

Gu, H., & Chen, X. (2011). Sacred groves sustain bio-cultural richness in yunnan stone forest. https://ourworld.unu.edu/en/sacred-groves-in-yunnans-stone-forest (retrieved on 10 February 2022).

Guha, R. (2000) *Environmentalism: A global history*. Longman. http://natureconservation.in/sacred-groves-of-meghalaya-complete-detail-updated/ (retrieved on 09 February 2022).

Jagdale, P. E. (2021). Role of sacred groves in conservation of biodiversity. *Journal of Emerging Technologies and Innovative Research*, *8*(9), c721–c728.

Jeeva, S., Mishra, B. P., Venugopal, N., Kharlukhi, L., & Laloo, R. C. (2006). Traditional knowledge and biodiversity conservation in the sacred groves of Meghalaya. *Indian Journal of Traditional Knowledge*, *5*(4), 563–568.

Kent, E. F. (2013). *Sacred groves and local gods: Religion and environmentalism in South India*. Oxford University Press.

Khan, M. L. (2022). Sacred groves in India. https://www.fao.org/3/XII/0509-A1.htm (retrieved on 10 February 2022).

Kharkongor, B., & Tiwari, B. K. (2016). Beliefs and institutional mechanisms for maintenance of sacred groves of Khasi Hills, Meghalaya. *International Journal of Current Research*, *8*(11), 42118–42123.

Komatsu, H., Rappleye, J., & Silova, I. (2019). Culture and the independent self: Obstacles to environmental sustainability? *Anthropocene*, *26*, 100198.

Laskar, B. I. (2021). *Sociology of Northeast India*. Rawat Publishers.

Malhotra, K. C. (1998) Anthropological dimensions of sacred groves in India: An overview. In P. S. Ramakrishnan, K. G. Saxena, & U. M. Chandrashekara (Eds.), *Conserving the sacred: For biodiversity management* (pp. 423–438). Oxford & IBH.

Mallarach, J. M. (Ed.) (2007). Nature and spirituality. *Proceedings of the 1st Workshop of the IUCN WCPA Delos.*

Maloney, M. P., & Ward, M. P. (1973). Ecology: Let's hear it from the people – An objective scale for measurement of ecological attitudes and knowledge. *American Psychologist*, *28*, 583–586.

MBB & WII (2017). *The Meghalaya state biodiversity strategy and action plan (Draft)*. Meghalaya Biodiversity Board & Wildlife Institute of India.

McLeod, C. (2002) *In the Light of Reverence*. Sacred Land Film Project, Earth Island Institute.

Mishra, B. P., Tripathi, O. P., Tripathi, R. S., & Pandey, H. N. (2004). Effects of anthropogenic disturbance on plant diversity and community structure of a sacred grove in Meghalaya, Northeast India. *Biodiversity and Conservation*, *13*, 421–436.https://doi.org/10.1023/B:BIOC.0000006509.31571.a0

Ormsby, A. A., & Bhagwat, S. A. (2010). *Sacred forests of India: A strong tradition of community-based natural resource management*. Cambridge University Press.

Oskamp, S. (2000). A sustainable future for humanity? How can psychology help? *American Psychologist*, *55*, 496–508.

Oviedo, G., Jeanrenaud, S., & Otegui, M. (2005). Protecting sacred natural sites of Indigenous and traditional peoples: An IUCN Perspective. *Gland*. https://www.iucn.org/sites/dev/files/import/downloads/sp_protecting_sacred_natural_sites_indigenous.pdf (retrieved on 06 June 2022).

Population census 2011. (n.d.). Census 2011 India. https://www.census2011.co.in/

Ray, R., Chandran, M. D. S., & Ramachandra, T. V. (2022) Sacred groves: Nature conservation tradition of the ancient world. http://wgbis.ces.iisc.ernet.in/biodiversity/sahyadri_enews/newsletter/issue32/article1.htm (retrieved on 10 February 2022).

Sah, P. (2022). Conservation of sacred groves of India: A traditional method for sustainability and environmental protection – A review. *Book of Abstracts: Indian Anthropology Congress -2022*, Indian National Confederation and Academy of Anthropologists (INCAA).

Sarkar, S. (2005) *Biodiversity and environmental philosophy: An introduction*. Cambridge University Press. http://doi.org/10.1017/CBO9780511498558

Sharma, S., & Devi, R. (2014). Sacred groves: Traditional way of conserving plant diversity in Block Bhalwal of Jammu District (J&K), *International Journal of Science and Research*, *3*(2), 239–242.

Singer, P. (2002). *One World*. Yale University Press.

Singh, S., Youssouf, M., Malik, Z. A., & Bussmann, R. W. (2017). Sacred groves: Myths, beliefs, and biodiversity conservation—A case study from Western Himalaya, India. *International Journal of Ecology*, *2017*, Article ID 3828609, 12 pages.

Vincent, M. K. K., Valdier, K. M., Benoit, O., Anderson, B., & Seri, G. B. (2019). Socio-anthropological factors as sources of desecration of Grébouo 1 sacred grove, Southwestern Cote d'Ivoire. *European Scientific Journal*, *15*(5), 178–195.

http://old.forest.kerala.gov.in/index.php?option=com_content&view=article&id=75:sacred-groves&catid=67:flora-and-fauna&Itemid=216 (retrieved on 10 February 2022).

http://www.cpreecenvis.nic.in/Database/Arunachal_Pradesh_878.aspx (retrieved on 09 February 2022).

http://www.cpreecenvis.nic.in/Database/Assam_2251.aspx (retrieved on 09 February 2022). http://cpreecenvis.nic.in/Database/Groves_811.aspx

http://www.megforest.gov.in/forest_sacredgroves.html (retrieved on 09 February 2022).
https://megbiodiversity.nic.in/sacred-groves (retrieved on 09 February 2022).
https://megforest.gov.in/forest_cover.html (retrieved on 14 July 2022).
https://nenow.in/NorthEast-news/sacred-forest-its-role-on-environment.html (retrieved on 09 February 2022).
https://www.aerfindia.org/grove.html (retrieved on 10 February 2022).

11 Nature-Based Livelihood and Economy

Tripura

Biswendu Bhattacharjee

The interaction of human beings and nature is an integral part of the evolution of civilization. The earliest human society appeared between river valleys and forests. Human beings survived by fighting with animals for food and shelter. Gradually, farming in the forest turned to developed agriculture, which became more sophisticated over millions of years, and now genetically engineered agriculture contributes to the survival of the growing population across the world. Agriculture-based human society steadily turned into an industrial society that has taken civilization to new heights.

Tripura is claimed to be one of the oldest princely States of ancient India. The princely rulers of Tripura claimed to have been descended from the Yoyati of the Lunar Dynasty of the Mahabharata. King Yoyati's successors were king Druhya, king Bavru, king Pratardan, king Daitya, king Tripur, and king Trilochana. Rajratnakar, the oldest State annals, endorses that king Bavru conquered the countries up to river Baitarini in Orissa, adding a portion of Burma to his kingdom, also bringing the Bay of Bengal under his control. King Bavru also attended the Ashvamedha Yajna (Horse Sacrifice) of Emperor Dasharatha of Ayodhya. King Tripur was a contemporary of king Yudhisthira of the Mahabharata era. King Trilochana is said to have attended the Rajsurya Yajna of king Yudhisthira. King Trilochana was a pious king, a votary of Lord Siva. The historical chapter of the royal line of Tripura known as the Manikya Dynasty had begun with the reign of Maharaja Maha Manikya who ascended the crown in 1400 AD Maharaja Maha Manikya was the first ruler who started ruling the State under the royal title of Manikya. In between the first ruler Maharaja Maha Manikya, and the last ruler, Maharaja Bir Bikram Kishore Manikya Bahadur, the princely State was ruled by the Manikya. For a long period of over 547 years altogether 35 Maharajas ruled the kingdom of Tripura. The last king Maharaja Bir Bikram Kishore Manikya Bahadur reigned for the period from 1923 to 1947. He was a versatile as well as a farsighted king who died on 17 May 1947. The economy during the entire history of the royal dynasty was mainly the revenue collection from the hill Tipperah where people practiced unsettled agriculture across the hills without having any rights to the land, which is known as Jhum (shifting cultivation).

DOI: 10.4324/9781003504870-13

Traditional Agriculture

Traditionally, shifting cultivation has been a practice in sub-humid and humid tropical hilly terrains where tribals have had the right to land either individually or on a community basis. Shifting cultivation has evolved in a historical context. It survives because the shifting cultivators have not been provided sufficient land where they can carry out a more settled way of cultivation. The natural process of evolution has been obstructed through encroachments where the space for settling down has been robbed for historical reasons. Shifting cultivation (slash and burn cultivation) is one major human impact on the environment in the tropics leading to habitat fragmentation and replacement of primary forests by secondary forests. It is practiced throughout the tropics and sub-tropics. It is one of the oldest cultivation systems, the origin of which can be traced back to the Neolithic period (13,000 to 3000 BC). About 40 million km of land (30% of the global arable land), inhabited by over 250 million people (7% of the world's population) mostly concentrated in Central and South America, the African tropics, and South-East Asian countries, is estimated to be under shifting cultivation. Shifting cultivation is popularly known as Milpa in Central America, Zande in Africa, Chena in Ceylon, and Caingin in Philippines. It differs from the advanced 'settled cultivation' in the shifting of the 'fields' rather than the 'crops' (Gupta, 2005). In India, around 2 million tribal people are engaged in shifting cultivation in 15 states, and approximately 11 million ha of forest land are affected by this practice in the states of the North East India, Orissa, Tamil Nadu, Sikkim, Maharashtra, and Karnataka to a lesser extent.

The North Eastern Hill Region of India contiguous with Montane Mainland of Southeast Asia covering parts of Cambodia, China, Laos, Myanmar, Thailand, and Vietnam is a region distinguished globally for a huge biological and cultural diversity and ecosystem services on one hand and marginality in terms of conventional socio-economic indicators of development on the other. Shifting agriculture, the major land use and backbone of rural economy, has been a focal point of development policies and programs (Borah et al., 2022). In the hilly region of North East India, shifting cultivation is a widely practiced food production system affecting an area of 2,56,083 sq km. This accounts for 7.76% of the country's total geographical area. Altogether 33 districts are affected by shifting cultivation in the North East region of India. Over a 100 tribal minority ethnic groups of Tibeto-Mongoloid origin are involved in shifting cultivation.

At least 100 different indigenous tribes and over 620,000 families in the seven states of North East India depend on Jhum for their subsistence (Ramakrishnan, 1992). The North Eastern region of India enjoys a rich natural resource base and biodiversity hotspots compared to other parts of the country. However, the region is not being systematically harnessed to generate wealth, because of which the region is considered backward. Indeed, many of the communities, particularly those living in the upland areas, continue to suffer from poverty and hunger. Shifting cultivation remains one of their main livelihood occupations.

History of Forest Dependency

Tripura is a remote, mountainous state in North East India, bordering with Bangladesh on three sides. The population is about 4.1 million (Estimated in 2019–20), of which 31% are a variety of ethnic minorities and the remaining are Bengali. Around three lakh people (living in around 400 villages) are dependent on forests for their livelihoods, which are mostly subsistence based. There is a common understanding that forest-dwelling communities are largely practicing shifting cultivation (Jhum) which has been the predominant land use in the hills of Tripura for a long time. However, the latest statistics revealed that hardcore Jhumia families in Tripura have been reduced to 6,979, which was over 25,000 a decade ago. Reduction of Jhum practitioners in no way has cut down the dependency on forests, rather it has increased in manifolds with conservation approaches. After the implementation of the Scheduled Tribes and Other Traditional Forest Dwellers (RoFR) Act, 2006, so far 130,901 families have been vested with the right to use forest land for livelihood. They are largely dependent on the forest ecosystems for their economy and lifestyle. Their economy was mainly based on food gathering, food production, arts and handicrafts, making articles and tools of daily use, and a barter system for exchange with other family units. The economy was simple, self-contained, and self-sufficient. There was a positive balance between their needs and the quantity produced or gathered. Their village organizations and social systems were used to ensure proportional distribution of land among families differing in population and economic status. An interview with the former minister of forests, Mr. Mevar Kumar Jamatia implied that any religious and spiritual customs at individual and collective levels were aimed at ensuring a conservative and respectful approach towards their major economic resource, that is, the forests.

Shifting cultivation is a unique feature of agriculture in the hilly region of the North East. Although this practice is criticized due to low productivity and environmental diseconomies the continuance of shifting cultivation is closely linked to the ecological, socioeconomic, cultural identity, and land tenure systems of tribal communities. It is worthwhile to mention that the strategies for improvisation of agricultural practices through modern technological interventions should be always in tune with the socio-cultural milieu of the local people of the North Eastern region to achieve definite benefits for socio-economic upliftment. Transfer of technologies for farm machinery, equipment, and even for machine/tools of production lines raises many adverse issues on farmers' safety, comfort, working efficiency, and so on but are unfortunately very often overlooked. Application of ergonomic knowledge from various sub-disciplines of ergonomics, for example agricultural ergonomics, occupational health and safety, environmental ergonomics, and design ergonomics might be useful for developing sustainable agricultural practices with better productivity and farmers' well-being.

Traditional agriculture is often considered a step between the local hunt-and-gather practice, which provides communities with subsistence levels of food, and the practices of modern agriculture, used for mass production of food for global distribution. The traditional agriculture practice develops a balance between meeting present needs, conserving natural resources, and protecting the environment for the benefit of future generations. Traditional agricultural approaches are not practical for mass food production but account for a substantial amount of local food production in the developing world.

In recent years, documenting traditional wisdom has gained significant attention the world over, because of its importance in developing a high-potential environment and sustainable management. Identification and utilization of such indigenous knowledge from the elderly people of rural sites and tribals will surely bridge the gap between current science and the age-old practices. Indigenous knowledge linked with the manipulation and use of natural resources in various ways forms the basis for their link with nature, and the varying levels of refinement depend on the level at which the society finds itself in the social evolutionary process.

The people of North East India represent a fascinating variety of cultures. Jhum plays an important cultural role in local customs, besides ensuring agro-biodiversity conservation and offering livelihood security to the rural upland poor. Shifting cultivation, locally known as Jhum, is a widely distributed form of agriculture in the upland areas of North East India. The practice involves site selection and slash and burn, followed by mixed cropping for a year or two and fallowing for certain years for recuperation of the land. Despite various efforts of the government to address and contain Jhum, over 443,000 families of upland rural communities of the region continue to be partially or wholly dependent on it for their livelihoods, with total areas affected by the practice estimated between 1.73 to 13.81 million ha (NEC, 2006).

Jhum in North East India, including its variants such as alder-based Jhum in Nagaland and bun cultivation in Meghalaya continue to attract diverse opinions. Its critics call it an inefficient and wasteful form of agriculture, while others see this as a diversified livelihood system that ensures sustenance along with the conservation of associated rich cultural heritage. The shortening of the Jhum cycle (the intervening period between fallowing and returning to the same spot for cultivation) from traditional 10 years or more to 4.5 years on average now is indeed a matter of concern. An article by Darlong (2015) indicated that this is seriously impacting the local livelihoods and environmental security in many pockets of the region. The superiority of Jhum cultivation over some forms of sedentary cultivation partly explains the persistence of this form of agriculture in North East India. Other reasons include the economic security provided by Jhum and its cultural importance to indigenous tribes. Poor access to markets, capital, and technological know-how of more commercially rewarding alternatives such as horticulture and cash crop cultivation also hinder the transition to other occupations. Clearly, one cannot do away with Jhum assuming it to be a primitive and inefficient system, as attempted in governmental Jhum control programs and new land use policies (Malik, 2003). Instead, an unbiased understanding of the advantages of Jhum is required for the proper design and implementation of development programs.

Jhum plays an important cultural role in local customs, traditions, and practices, besides offering economic security to farmers. In fact, Jhumias are a tribe who practice shifting cultivation or jhuming. Over the years the Jhum economy has undergone many changes – land available for Jhuming has decreased, leading to a shortening of the Jhum cycle and a fall in incomes. Not only has the plight of the existing Jhumias deteriorated, but a new group of Jhumias has emerged due to the several 'development' programmes whose plight is even worse. The government has made some efforts at resolving the problems of the Jhumias (Dasgupta, 1986). It would be unfortunate if developmental programs based on misguided opinions about Jhum suppress this unique form of agriculture. Only occupations providing monetary and social benefits embodied by Jhum are likely to gain acceptance. A balanced approach to development that also recognizes the merits of Jhum is needed. It is claimed that 1.6 million person-years of employment in India are from non-timber forest products (NTFP) while the forestry sector in total provides 2.3 million person-years of employment (Shiva & Mathur, 1996. However, indiscriminate forest loss for development purposes has become a challenge for monetizing nature-based economies across the country. The Forest Survey of India report says India lost over 1.6 million ha of tree cover between 2001 and 2018. The annual forest loss in the year 2016 was 3,310 ha, followed by 3,060 ha in 2017 and 3,537 ha in 2018. The numbers have soared drastically, as it has generally remained below 3,000 ha, except in 2007, when it was 4,145 ha. The five North Eastern states – Nagaland, Tripura, Meghalaya, Mizoram, and Manipur – were responsible for over 50% of all tree cover loss in the same period, as per data analyzed by the Forest Survey of India. The loss of tree cover contributed to 172 metric tons of carbon emissions in India during this period.

Dependency on Nature

Development activities and forest fires are the main causes of the decline of forest cover. The Botanical Survey of India (BSI) has found out in a recent study that 13 native trees are on the verge of extinction in the wild as reported by the national daily, *Hindustan Times*. The BSI classified these species as 'endangered' and 'critically endangered' between 2013 and 2017. The list of tree species includes Assam catkin yew, Indian bdellium tree, white cedar, and

Minangmose. They are spread across the North East, central, and southern parts of India.

The Department of Forest in Tripura in 2008 has undertaken a study to quantify the dependency of three major tribal communities of the state – Tripura, Reang, and Chakma. The final report revealed that except for Chakma, the other tribes were found to follow the Hindu religion. Christianity was also practiced by some members of the Reang community. With both Hinduism and Buddhism being oriental religions and being practiced in traditional forms, these communities are expected to have greater cultural livelihood linkages and use of biodiversity. The average family size is about four. Tripura tribes seem to have comparatively bigger families. About 9% of the households are controlled by women in this tribe and the Tripura youth was found to marry earlier than their counterparts, while marriage for Chakma seems to be at a later age. The sex ratio is not very well balanced with an average of 916 females for 1,000 males. There were a smaller number of women in the Tripura community, while the number was more visible in the Chakma community. Like other areas, these communities were found to go for nuclear families, and most of the respondents were married. Among the three tribal communities in Tripura, the Tripura seems to be using more species and Chakmas the least, with huge seasonal variations. The use of biodiversity is more during the autumn and winter seasons and vegetable items have maximum variety followed by fishes, green leaves, and meats. The report denotes S1 (Summer), S2 (Autumn), S3 (Winter), and S4 (Monsoon).

Table 11.1 Seasonal variation in forest produce collection NTFP (No. of species)

Forest items	*Tripura*				*Reang*				*Chakma*			
	S1	*S2*	*S3*	*S4*	*S1*	*S2*	*S3*	*S4*	*S1*	*S2*	*S3*	*S4*
Barks	1	2	2	1	4	0	0	0	0	8	0	0
Fibers	3	2	2	8	1	4	2	4	3	0	0	0
Fish	21	24	23	6	23	24	18	16	10	14	10	1
Flower	2	4	2	4	5	12	0	6	2	0	0	5
Fruits	29	17	13	1	13	8	7	13	2	23	10	6
Insects	7	6	4	2	2	4	0	0	4	6	1	4
Leaf	19	28	21	10	17	25	9	11	5	23	8	7
Meat	17	15	12	6	16	21	6	9	2	22	10	5
Roots	7	13	16	6	5	0	1	0	3	12	9	0
Seeds	5	13	14	3	5	0	3	3	0	9	8	8
Shoots	7	9	8	6	15	12	4	6	3	9	5	5
Vegetable	64	45	40	16	55	31	15	24	31	25	28	0
Gross Total	182	178	157	69	161	141	65	92	65	151	89	41

Source: Study Report on Socio-economic, Cultural & Traditional Aspects of Tribal Jhumia Families Belonging to Tripuri, Reang, and Chakma Communities, Department of Forest: 2008.

Table 11.2 Seasonal variation in forest produce collection (quantity/community wise)

Forest Produce	Tripura		Reang		Chakma	
	Kilogram	*Number*	*Kilogram*	*Number*	*Kilogram*	*Number*
Barks	178.76	457.19	110	1	103.54	0
Fibres	405.43	1,101.83	237.4	98.2	128.39	0
Firewood	1,593.06	2,145.12	803.8	258.7	210.49	10.56
Fish	105.90	810.98	52.9	1,401.3	107.92	0
Flower	117.80	1,064.02	55	209.4	12.04	0
Fruits	257.52	529.2	88	16.3	116.87	0.5
Insects	6.12	1,860.35	196.9	11.5	84.12	0.5
Leaf	185.30	1,125.84	69.8	533.7	117.17	8.8
Meat	167.60	86.1	121.7	118.4	117.49	0
Roots	283.68	22.2	254.2	91.1	145.73	0
Seeds	204.49	180	138.8	6	154.45	0
Shoots	558.85	349.78	176.7	93.9	117.7	4.92
Vegetable	240.09	356.53	240.5	113.3	169.84	0
Total	4,304.58	10,089.14	2,545.7	2,952.8	1,585.75	25.28

Source: Study Report on Socio-economic, Cultural & Traditional Aspects of Tribal Jhumia Families Belonging to Tripuri, Reang, and Chakma Communities, Department of Forest: 2008.

The report further stated that a Tripura household collects more forest produce, almost four times more than a Chakma household over four seasons. On average, a Jhumia household collects about 3 metric tons of forest produce per year, a major constituent of which is firewood, consisting of almost one-third of the produce. The major forest products in terms of the amount of collection per household are shoots (11%), roots (8%), and vegetables (7%). Fish, meat, and insects together constitute about 10% of the total biomass collected. Among Chakma households collections are almost evenly spread across different categories of forest products, while among the households of other tribes collections across categories are a little skewed in favor of firewood, vegetables, and leaves.

Economy of Nature

According to Biological Diversity Act (2002), priority issues were set up for sustainable use of biodiversity, access and benefit sharing of biodiversity for commercial use, identification of species of conservation importance, and documentation of Peoples Biodiversity Registers. In Tripura, to support rural livelihood, Tripura Biodiversity Board has shared the benefits with the communities while value-addition activities of NTFP have been continuing under the externally aided project of the Japan International Cooperation Agency (JICA).

Plants are natural industries that provide high-quality food and raw materials for the pharmaceutical, cosmetic, and perfumery industries. The reports of the side effects of synthetic drugs and fast food have necessitated the exploration of the efficacy of purely natural products. This has led to the revival of herbal treatments for many diseases. Of all the medicine in clinical use, there are about 30% of drugs derived from plants, and about 80% people in developing countries rely on traditional plant-based medicine. To cope with the increasing demands for food and medicine steps are being taken to explore and populate important economic plants with established food and medicinal efficacies. Traditionally, local communities worldwide are extremely knowledgeable about the local plant resources on which they are so intimately and immediately dependent. Unfortunately, much of this knowledge is lost today, as traditional cultures are getting eroded.

Among the strategies used by the rural people to overcome food shortage is the tendency to depend on wild food, including wild fruits. These edible plants may greatly contribute to the diet of rural people by providing rare nutrients and facilitating survival in a time of famine (Guy-Alain & Francois 2003. Such wild edible plants not only supplement the food quantity but make a significant contribution to the population's nutrition throughout the year (Ogle Britta et al. 2003). The use of many wild species by the tribals to meet their diverse requirements is largely due to the prevalence of diversity of vegetation in an area (Jha, 2015).

There are about 19 ethnic groups predominantly living in and around the dense forests of Tripura, cooperating with nature, with their own language, culture, food habits, and socio-religious traditions. The sustainable utilization of wild plants in daily life for food, medicine, fodder, house building, craft, and agriculture is an integral part of rich tribal culture and tradition. The collection of wild plant parts for a livelihood by the forest dwellers is not only to sustain their socio-economic and cultural life, but these types of minor products play a vital role in supplementing the daily diet rather than generating small amounts of income by selling these in village markets. Particularly, the availability of some seasonal wild plants in the forest around their habitation not only enhances their annual food storage or capital gain, but it is the traditional practice of seasonal gathering of nutrition and medicine. Considering this fact, an attempt has been made to document the wild edible plant resources, especially those which are consumed.

There are 40–100,000 plant species that have been regularly used for food, fiber, shelter, industrial, cultural, and medicinal purposes. However, only a small number of plants are widely used. The remaining plant diversity is underutilized in most tropical countries. Underutilized plants contribute immensely to family food security and serve as a means of survival during times of drought, famine, shocks, and risks. Wild edible fruit-yielding plants are known to supplement nutritional requirements due to their better nutritional value. Tripura Biodiversity Board has so far listed the following plant species, which are edible not only as food but also have medicinal values.

Table 11.3 List of wild edible vegetables traditionally consumed by the tribes of Tripura

SL	*Name of the plants*	*Family*	*Local Name*	*Parts used*
1	Amorphophallus bulbilfer (Roxb) Bl.	Araceae	Dangadonga	Stem, Rhizome
2	Alpinia allughas (Retz.) Roscoe	Zingiberaceae	Tara	Stem
3	Amaranthus spinosus L.	Amarenthaceae	Katamairah	Tender shoot
4	Brassaiopsis griffithii C.B. Clarke	Araliaceae	Chapok	Flower
5	Bambusatulda Roxb.	Poaceae	Mritinga	Shoot
6	Coccinia grandis (L.) Voigt	Cucurbitaceae	Talakachu	Leaf
7	Colocasia esculenta (L.) Schott	Araceae	Kacu	Whole plant
8	Canavalia gladiata (Jacq.) DC	Papilionaceae	Makhan Sim	Pod
9	Dendrocalamuslongispathus Kurz	Poaceae	Rupai	Shoot
10	Dioscoreabulbilfera L.	Dioscoreaceae	Thaborchok	Tuber
11	Dioscoreahamiltonii H. K	Dioscoreaceae	Thanara	Tuber
12	Diplazium esculentum (Retz.) Sw.	Athyriaceae	Dhekisag	Shoots
13	Euryle ferox Salisb.	Nymphaeaceae	Makhna	Fruit
14	Homalomenaaromatica Schott.	Araceae	Gandhauri	Leaf, rhizome
15	Hodgsonia macrocarpa (Bl). Cogn	Cucurbitaceae	Bonkomar	Seed
16	Ipomoea aquatica Forsk.	Convolvulaceae	Kamli Sag	Leafy stems
17	Leucas aspera (Wild.) Link	Lamiaceae	Dron	Leafy stems
18	Lasia spinosa (L.) Thw.	Araceae	Kaitaish	Stem
19	Musa acuminata Colla Musaceae	Bonkala	Inflorescence,	Xylem
20	MelocanabacciferaTrin.	Poaceae	Muli	Shoot
21	Monochoria hastate (L.) Solms	Pontederiaceae	Chichiri	Whole plant
22	Momordica cochinchinensis (Lour.) Spreng.	Cucurbitaceae	Bonkakrol	Fruit
23	Nymphaea nouchaliBurm. f.	Nymphaeaceae	Sefla	Stem with flower
24	Neptunia prostrata (Lamk.) Baill	Fabaceae	Khorai	Whole plant
25	Ocimumamericanum L.	Lamiaceae	Banta	Leaf
26	Oroxylum indicum Vent.	Bignoniaceae	Thana	Pod
27	Polycarponprostratum (Forssk.)	Caryophylaceae	Gima	Whole plant
28	Solanum torvum Sw.	Solanaceae	Titbegun	Fruits
29	Solanum surattenseBurm	Solanaceae	Khamka	Fruits
30	Zanthoxylum limonella (Dennst) Alston	Rutaceae	Zarai	Leaves

Source: Technical Report Documentation of Some Indigenous Traditional Knowledge (TK) and their Prioritization for Intellectual Property Rights (IPRs) issues in Tripura, Tripura Biodiversity Board, 2016.

Case Study

Thirty-six-year-old traditional healer of Hazaripara village of Rajnagar under Khowai police station, Santi Kumar Debbarma, revealed that even after the advancement of medical science and extension of healthcare facilities, many people in remote parts of Tripura are still visiting traditional healers for physical problems. This trend is not only restricted to the tribals or rural people. People from various walks of life across communities come to traditional healers with a belief that the nature-based traditional treatment can give relief for the most complicated diseases. According to Debbarma, the traditional healers of his circle are experts in treating bone fractures, joint pain, snakebite, skin disease, gynecological ailments, mental disorders, gall bladder stones, malaria, jaundice, pneumonia, leucorrhea, and ailments of eyes and ears with specific plant parts, which are abandoned in the forest. He specifies some plants having medicinal values that are used by the traditional healers of Tripura for treatments.

- Name of the Plants : Bon Alach
  Scientific Name : Amomum ligniform (Roxb.)
  Availability status : Wild Rare
  Part(s) Used : Rhizome
  Name of the Ailments : Muscular rheumatism
  Mode of Administration : The dried rhizome is powdered, and half a teaspoonful of powder is mixed with an equal amount of honey and taken once a day for about one month in case of muscular rheumatism.
- Name of the Plants : Agar
  Scientific Name : Aquilaria malaccensis Lamk
  Availability status : Rare
  Part(s) Used : Bark
  Name of the Ailments : Leucodermic and rheumatism
  Mode of Administration : Poultice of bark is given to cure rheumatic pain, and decoction of the bark is rubbed on the spotted areas before bathing till cured.
- Name of the Plants : Satamuli
  Scientific Name : Asparagus racemosus Wild.
  Availability status : Wild
  Part(s) Used : Tuberous roots
  Name of the Ailments : Epilepsy
  Mode of Administration : Half a cup of tuberous root's decoction diluted with an equal amount of milk is taken once a day for three months as a remedy against epilepsy.
- Name of the Plants : Ban Kalai
  Scientific Name : Atylosia scarabaeoides (L) Benth
  Availability status : Common (wild)
  Part(s) Used : Leaves and seeds
  Name of the Ailments : Skin disease
  Mode of Administration : Crushed leaves and seeds are applied externally on the body for any kind of skin disease in children.

- Name of the Plants : Ziodi
  Scientific Name : Barleria prionitis L.
  Availability status : Common (wild)
  Part(s) Used : Leaves
  Name of the Ailments : Pruritis, rubefacient, and blotch.
  Mode of Administration : Several leaves mixed with the ashes of Terminalia chebula fruit and sulphur dioxide (Gandhak) are crushed and made into pills (500 mg) and then dried. About 1–2 pills are mixed with 100 ml of coconut oil and massaged every day all over the body.
- Name of the Plants : Hijol
  Scientific Name : Barringtonia acutangula (L) Gaertn.
  Availability status : Occasional (wild)
  Part(s) Used : Bark and tender leaves
  Name of the Ailments : Worm infection, boils.
  Mode of Administration : Decoction of bark about 5ml, used twice a day for 2–3 days in worm infection; juice of tender leaves is applied in boils to promote suppuration.
- Name of the Plants : Chapok (K)
  Scientific Name : Brassaiopsis glomerulata (Bl.) Regel
  Availability status : Occasional (wild)
  Part(s) Used : Wildflower bud
  Name of the Ailments : Gastritis, ulcer, and jaundice
  Mode of Administration : The immature flower is cooked with common vegetables and taken with rice during severe gastritis and ulcer, juice is extracted from both mature and immature flowers and prescribed 2 cups a day during jaundice.

Contribution of Nature to Economy

The nature-based economy, as seen in Tripura, indicates that man and environment are interdependent; man being dependent for its survival. The cycle of degradation and restoration is bound to continue as man is dependent on the environment to make his ends meet.

Traditionally, although extracting natural resources from the forest did not have any control, after the Biological Diversity Act, 2002 was passed by Parliament restrictions have been imposed for harnessing minor forest produce. Under subsection (2) of section 3 of the Act, those who intend to have access to biological resources and/or associated traditional knowledge for research or bio-survey and bio-utilization for research shall apply to the National Biodiversity Authority (NBA) for obtaining access to such resources in India. The NBA shall, on being satisfied with the application under sub-regulation (1), enter into a benefit-sharing agreement with the applicant which shall be deemed as a grant of approval for access to the biological resource for research referred to in that sub-regulation. Provided that in the case of biological resources having high economic value, the agreement may contain a clause to the effect that the benefit-sharing shall include an upfront payment by the applicant, of such amount as agreed between NBA and the applicant. Any person who intends to have access to biological resources harvested by Joint Forest Management Committee/ Forest dweller/ Tribal cultivator/ Gram Sabha shall apply to the NBA or State Biodiversity Board. The benefit-sharing obligations on the trader shall be in the range of 1.0 to 3.0% of the purchase price of the biological resources and the benefit-sharing obligations on the manufacturer shall be in the range of 3.0 to 5.0% of the purchase price of the biological resources. Where the trader sells the biological resource purchased by him to another trader or manufacturer, the benefit-sharing obligation on the buyer, if he is a trader, shall range between 1.0 to 3.0% of the purchase price and between 3.0 to 5.0% if he is a manufacturer. Under this provision, Tripura has shared the benefits of biological resources with the community to the tune of Rs 34.66 Cr in the last three years which contributed to the nature-based economy of the state.

Developing a nature-based economic model, Tripura demanded a green bonus of Rs 2,125 Cr between 2015–16 and 2019–20 for maintaining a clean environment and it was not turned down by the 14th Financial Committee along with a grant of Rs 1,250 Cr for climate change mitigation and adaptations. Thereafter, aiming to increase carbon stock as per the commitment of India to the global community and to boost short-term–forestry-based livelihood, Tripura brought Rs 1,000 Cr JICA–aided second phase project for sustainable catchment forest management. The project was designed with such components, which will not only improve the green cover and ecology but help forestry-based direct income generation activities for about two lakh families living across seven districts of Tripura until 2029–30 under the project. To improve forest quality it targeted all 11 major river catchments through sustainable forest management, soil and moisture conservation, livelihood development through value addition of non-timber forest produces, and eco-tourism under this externally aided project. The state government has brought a comprehensive model aiming to make Rs 2,000 Cr economy by 2025 through Agar cultivation, from which the Agartala (the capital of Tripura state) name was derived.

In Conclusion

As cited Tripura has a major dependence on nature and its products. Sustainable practices have the capacity to drive innovation and develop new technologies, bring about new products, and mobilize services. This in turn leads to economic growth by creating new job markets. There has been an incredible shift in the market choices of consumers; people are now drawn towards plant-based organic products be it food, furniture, cosmetics, medicines, and many such consumables. This shift is a positive signal for economies like Tripura, which is largely nature dependent. If the graph continues to bend towards sustainable products, it indeed will serve as a game changer for the economy of Tripura. Moreover, preserving nature benefits mental and physical health in a positive way. The natural environment serves as an asset

Table 11.4 Access and benefit sharing FY 2019–20 to 2021–22 of forest products between Biodiversity Management Committee (BMC) and Tripura Biodiversity Board (TBB)

Year	*Forest Product*	*Collection*	*Sale (Rs)*	*Income (Rs)*	*BMC (2%)*	*TBB (1%)*	*BMCs benefitted*
2019–20	Broom Grass	1,019,750 kg	Fresh: 15–17/kg Dry: 45/kg	19,331,250	386,625	193,312	47
	Gandhaki	1,439,400 kg	Fresh: 25/kg Dry: 185/kg	35,342,000	706,840	353,420	43
	Bamboo for umbrella handle	2,650,000 Nos	1.50 to 2.00/piece	4,880,000	97,600	48,800	46
	Bamboo Stick for bag handle	530,000 Nos	1.50/piece	795,000	15,900	7,950	11
	Bangphai	1,000 kg	200/kg	200,000	4,000	2,000	1
	Forest cane	15,000 Nos	3/piece	45,000	900	450	1
	Muli & Mrittinga bamboo	1,200,000 Nos	Muli: 10/piece Mrittinga: 20/piece	170,000	34,000	17,000	4
	Boro Elaichi	15,000 kg	20/kg	300,000	6,000	3,000	2
	Black pepper	10,000 kg	50/kg	500,000	10,000	5,000	1
	Betu, Makal, Bari, Kanak Kaich, Rupai bamboos	1,028,177 Nos	40/piece 25/piece 35–60/piece 10/piece 7/piece respectively	11,573,905	—	578,697 *(5% of the total sale value)*	M/S Mutha Industries Pvt. Ltd, Bodhjungnagar
Total				74,667,155	1,261,865	1,209,629	157
2020–21	Broom Grass	578,000 kg	Fresh: 17–20/kg Dry: 45/kg	12,589,500	251,790	125,895	39
	Gandhaki	236,000 kg	25/kg	5,900,000	118,000	59,000	24
	Bamboo for umbrella handle & bag handle	1,266,550 Nos	1.50–2.00/piece	2,344,825	46,896	23,448	27
Total				2,083,4325	1,646,191	963,153	90

(Continued)

Table 11.4 (Continued)

Year	*Forest Product*	*Collection*	*Sale (Rs)*	*Income (Rs)*	*BMC (2%)*	*TBB (1%)*	*BMCs benefitted*
2021–22	Broom Grass	1,186,500 kg	Fresh: 17–20/kg Dry: 48/kg	2,719,500	543,900	271,950	58
	Gandhaki	789,500 kg	25/kg	19,737,500	394,750	197,375	26
	Bamboo for umbrella handle & bag handle	2,672,000 Nos	1.50–2.00/piece	5,344,000	106,880	53,440	26
Total				52,276,500	1,045,530	522,765	110

for goods and service production, and exposure to nature boosts human mental health and well-being. Poor mental health imposes major costs on economies, caused by poorer efficiency and productivity in the workplace (Buckley et al., 2019).

References

Biological Diversity Act. (2002). http://nbaindia.org/uploaded/docs/biological-diversityact-ii.pdf

Borah, J. R., Gilroy, J. J., Evans, K. L., & Edwards, D. P. (2022). The value of shifting cultivation for biodiversity in Northeast India. *Diversity and Distributions*, *28*(9), 1979–1992. https://www.jstor.org/stable/48682409

Buckley, R., Brough, P., Hague, L., Chauvenet, A., Fleming, C., Roche, E., Sofija, E., & Harris, N. (2019). Economic value of protected areas via visitor mental health. *Nature Communications*, *10*, 5005. https://doi.org/10.1038/s41467-019-12631-6

Darlong, B. (2015). *Sacred groves: A study of the religious beliefs of the darlong community in tripura*. https://chitrolekha.com/sacred-groves/

Dasgupta, M. (1986). Jhumias of tripura. *Economic and Political Weekly*, *21*(44/45), 1955–1960. http://www.jstor.org/stable/4376302

Gupta, V. (2005). Jhum cultivation practices of the Bangnis (Nishis) of Arunachal Pradesh, *Indian Journal ofTraditional Knowledge*, 4(1), 47–56.

Guy-Alain, A. & Francois, M. 2003. How Ivory Coast's Malinke ethnic group select most beneficial wild fruits. *Agroforestry Today* 13 (1–2): 2–6.

Jha, K. K. (2015). Non-timber forest products, their vulnerability and conservation in a designated UNESCO Heritage Site of arunanchal pradesh, India. *Notulae Scientia Biologicae*, 7(4), 9701. https://doi.org/10.15835/nsb.7.4.9701

Malik, B. (2003). The "Problem" of shifting cultivation in the Garo Hills of North East India, 1860–1970. *Conservation and Society*, *1*(2), 287–315. http://www.jstor.org/stable/26396409

NEC. *Government of Assam Soil Conservation Directorate of Soil Conservation*. North Eastern Council (NEC) | Directorate of Soil Conservation | Government of Assam, India. (2006). https://soildirectorate.assam.gov.in/schemes/north-eastern-council-nec

Ogle Britta, M., Ho Thi Tuyet, Hoang Nghia Duyet and Nguyen Nhut Xuan Dung. Food, Feed or Medicine: The multiple functions of edible wild plants in Vietnam. *Economic Botany* 571: 103–117. 2003.

Ramakrishnan, P. S. (1992). *Shifting agriculture and sustainable development: An interdisciplinary study from North Eastern India. MAB series*, Volume 10, UNESCO, Paris.

Shiva, M.P. & Mathur, R.B. (1996), *Management of minor forest produce for sustainability*. Oxford and IBH Publishing Co. Pvt. Ltd., New Delhi.

12 Eco-Feminism in the Context of the Khasi of Meghalaya

Rekha M. Shangpliang

There has been much exploration of the belief that human beings differ from animals as much as the corollary that 'nature' is distinct and contrasted with 'culture'. The historical shaping of the relationship between nature and culture clearly spells out the sharp contrasting image between the two. There is a metaphoric transformation inherent in what we call the 'Nature-Culture dualism'. Structuralists such as Levi-Strauss (1967) attribute this transition as a product of the natural process of evolution; when humans started becoming humanlike, they created culture by using symbols such as kinship and language. The idea of equating 'woman' with 'nature' has a long-standing history rooted in the beginning of civilization when the hierarchical structure of society (based on class and gender) was clearly established and became part of the natural world.

Women have long been associated with nature. Judith Plant (1992), in the book *The Green Reader*, supports this statement by citing examples of the use of certain metaphors in our everyday language, for example, the phrase 'a virgin forest' is often used to mean a forest that is devoid of exploitation and untouched by man. The idea of looking at the earth as a parental figure that sustains mankind, that acts as a life-giving force, has been the root of many traditions and belief systems all over the world. It is believed that the first Greek God was the Goddess 'Gaia' or 'Mother Earth' who was created out of primordial chaos, bringing out her feminine elements of fertility, wholeness, balance, and continuity of life. From her fertile womb all life sprang, and unto the bosom of Mother Earth all creatures on earth live in harmony until the span of life ends. Being metaphorically explained as 'Mother Earth', Greek mythology offers us a platform for understanding the beginning of such a belief which is deeply rooted in many cultures of the world.

Such historical connections between women and nature gave birth to a series of movements in the 70s that began as environmental movements. The feminists who then took interest in such movements tried to explore how oppressions were linked through gender, race, and ecology. This ultimately led to a new wave of 'eco-feminism,' a term originally conceived by Francoise d'Eaubonne in 1974, which was a philosophy that emphasized the ways in

DOI: 10.4324/9781003504870-14

which both nature and women are treated by a patriarchal or male-centered society thereby advocating ways in which women must work to end the domination on nature if they were to work towards their own liberation. Women were therefore seen as the repository of the natural laws, the natural morality, which allowed them to sink within the wilderness of nature. But the link between nature and women is not 'given' (Mac Cormack & Strathern, 1980). The nature – culture – gender thesis is rooted in the ideology of linking femaleness to biology and maleness to society (Beauvoir, 1949), which is a construct of culture rather than a fact of nature.

Much of ecological history has tried to unfold the origin of the ideology of domination rooted in patriarchy, which was the underlying principle of man's control over nature. Based on the search for the origin of such an ideology, that is, equating women with nature and man controlling both – nature and women – Maria Mies, a well-known eco-feminist, traces the development of men's domination over women and nature through a materialistic dimension of the male body. Her main thesis is that the male body must mediate with external nature (through tools) to feel the productivity of its own body, as a human, in other words, a productive being (Datar, 2011). Thus, 'humanness' is identified with 'maleness' which is the very basis of the famous adage 'woman-the-gatherer' and 'man-the-hunter'. Historically, and across cultures, women have been perceived as closer to nature because of their physiology, social roles, and psychology. Such perceived notions have pushed women to the domestic sphere.

Eco-Feminism and the Indian Woman

The history of almost all cultures is replete with the identification and association of women and nature. This nature-feminine principle has been a dominant theme of a new era of 'eco-feminism', which is concerned with the broad contours of social construction and interpretation of women's relationship with the environment.

In India, the role of women in ecology movements can be traced back to the Chipko Movement, which was concerned with the preservation of forests and maintenance of the traditional eco-balance in the sub-Himalayan region. To many people, 'Chipko' evokes an idea of the most influential environmental movement in history. What happened 51 years ago in 1973, in a small hamlet in Chamoli district of Garhwal, when women activists staged protests against tree felling by literally 'hugging' the trees have today become the starting point of a conservation endeavor that was firstly 'pro-poor' and above all 'women-centered'. What is remarkable is the fact that Chipko has grown from being a 'forest-based' movement to becoming an environmental icon to the world today that has inspired 'eco-feminism' with a strong message – 'our lives before our trees' (Shangpliang, 2016). However, it is not enough to conclude that the Chipko movement is a celebrated environmental movement today. Although this movement has attained international recognition for a so-called

Third World country, there still seems to be very little hope for the problems of rural women. Looking at the daily lives of village women in our country the question remains – How do we generalize about women's ecological practices to the point that their roles become recognizable without sacrificing the enormous diversity of their lives?

Given such a conceptualization of women and nature, and the status of women in India vis à vis their various roles which set them apart from the men, the situation of Meghalaya is no different, yet it stands out due to its cultural ethos. There is a close symbiosis between nature and the daily life and sustenance of the people of Meghalaya and the Khasi community. The 'Nature-God-Human' principle which is so deeply rooted in the traditional Khasi faith leads one to understand that there is a close connection between the cultural and biological diversity amongst them. Women, being central to the Khasi matrilineal structure, have occupied an important place in the society, culture, and economy of the Khasi. Their gendered roles have centered around the idea of attributing 'motherly' or 'feminine' characteristics to nature.

The Khasi and the Natural Environment

The Khasi are a matrilineal tribe inhabiting the Khasi Hills in the North Eastern state of Meghalaya in India. Meghalaya, a Sanskrit word which means 'abode of clouds', evokes a unique feeling of serenity among artists who have drawn inspiration to paint many a dramatic canvas of its natural beauty. Perhaps it was this unique blend of beauty amidst simplicity that inspired the hill leaders of this region to fight for a separate state and thereby paved the way for a geo-political history that marked the emergence of Meghalaya as an Autonomous State on 2 April 1970 and as a full-fledged State on 21 January 1972.

Meghalaya is a hilly region bounded by Bangladesh on both the southern and western fringes and to the north by Goalpara, Kamrup, and Nagaon Districts of the Assam valley and to the east by the Mikir Hills and North Cachar Hills. Meghalaya is the only state in the North East with all three resident communities – Khasi, Garo, and Jaintia – following the matrilineal system. Matriliny has been the cradle of kinship system in the state with the prominent roles of 'Ka Khadduh' (youngest daughter) and 'Nokna' (Garo heiress) still holding an important place in their matrilineal structure.

While there are many interpretations of the origin of the Khasi race, the term 'Khasi' is believed to have derived its meaning from the terms '*kha*' meaning 'born of' and '*si*' meaning 'ancient mother', which when interpreted means 'born of the mother'; thus, bringing out the matrilineal character of the Khasi who trace their descent from the mother. The Khasi follow the principles of matrilineal descent where ancestral property is passed from the mother to the daughter, preferably the youngest daughter or the '*khadduh*'. The matrilineal Khasis of Meghalaya believe in the ideology of '*Long jaid naka kynthei*' which means 'From the woman sprang the clan'. This ideology is so deeply rooted in the Khasi ethos that it has brought to light the role of women in perpetuating the clan or '*kur*' because the children belong to the descent group of the mother. Therefore, it is customary for them to speak of a family of brothers and sisters who are great-grandchildren of one great-grandmother and identify themselves as '*shi-kpoh*', which literally means 'one-womb' or 'the issue of one womb'. Significantly, the Khasi ideology of human reproduction describes the father as the provider of stature and form (*U kpa uba I ka long rynnieng*), while the mother contributes flesh and blood (*Ka kmie kaba ai iaka doh ka snam*) to the child. This filial bond between the mother and the child is strongly reflected even in familial relations and kinship ties where the term *parakur* (mother's kins) is used more significantly than the word *bakha* (father's kins).

A woman is central to the family in Khasi ideology. She has the responsibility of nurturing and continuing the matrilineal descent line through reproduction. The Khasi woman has numerous familial and kinship roles to perform. Besides being the custodian of family property, the '*khadduh*' or youngest daughter's role is all the heavier and more burdensome. She must look after the aged members of the family, and other members of the matricin; she inherits the family property and continues the tradition of carrying out religious rites and ceremonies from time to time, although with the help of her mother's male relatives. The Khasi belief in the saying '*ka kynthei ka bat ia ka niam*', meaning 'It is the woman who holds the religion' signifies her responsibilities in the religious domains.

However, the exclusion of Khasi women from having any say in the village *durbars* and meetings is noteworthy. She is not allowed to speak aloud in public meetings and village council meetings perhaps because of her domesticated role and nurturing skills for which she is expected to confine her duties to the home and family matters, leaving the others to the male. Right from the time of recorded history Khasi women were known to have freely participated in economic activities outside home. They have had the freedom of choice in matters relating to marriage, household matters, education, and so on. As pointed out by Gurdon (1975), "Khasis carry very heavy burdens on their back. Women carry almost as heavy loads as men. They are certainly more industrious than the Assamese. Khasis are devoted to their offspring and the women make excellent nurses for European children."

The Khasi have a deep-rooted affinity with nature. The earth, which is symbolized as 'mother' or '*Mei Ramew*', is looked upon as the caretaker of all the natural resources. Thus, the forest holds an important place in the socio-economic and religious life of the Khasi. It is looked upon as a well-loved home, a game sanctuary, and an abode of worship, all rolled into one, around which their social, cultural, and religious activities revolve. It is a familiar sight to see women and children setting off into the woods to collect edible roots and fruits. A typical day out for an average Khasi village woman would be best described as follows:

> She carries her baby on her back, climbing the hills, she sets off for the woods with a ja-song (cooked rice packed in a leaf) some salt and dry fish (ktung). Going to the forest virtually takes her the whole day. So besides collecting twigs and fuel wood and other forest products, she spends the whole day washing clothes on the riverside in the foothills only to return home at dusk just before the sun sets with a head load of fallen twigs and firewood enough to keep the hearth burning for the night.
>
> (Shangpliang, 2013a)

These activities demonstrate the close engagement that Khasi women have with the forests and their environment and with nature in general. Over centuries the close symbiotic relationship between women and the natural environment has led to a repertoire of indigenous knowledge. Thus, understanding the critical role played by women in conservation of forests requires empowerment of women at the grassroot level giving them opportunities for decision-making in management of local forests where,

> The true aim of conservation, then, is twofold viz: to ensure the preservation of a quality environment that considers aesthetic and recreational as well as product needs and to ensure a continuous yield of useful plants, animals and materials by establishing a balanced cycle of harvest and renewal.
>
> (Raimi et al., 2019)

While agriculture is the mainstay of the people of this state, especially in the rural areas, the role of women in natural resource management practices has played a major role in nurturing 'Mother Earth' as 'Mei-Ramew'. However, the plight of women in economic pursuits in the state, let alone livelihood strategies by rural women, is still a neglected area. Moreover, many women in the state bear the burden of single-headed households with poor health care facilities. Maternal and child health care is also an area that needs attention. Sex ratio is an indicator of the health, nutrition, and survival status of women. A low gender ratio (986 per 1,000 males in the 2011 Census) means both a lower social status of women and discrimination at various levels, especially inside the home. As per the provisional population results of the 2011 Census, the overall sex ratio (total population at all ages) at the national level is 940 females for 1,000 males. Among the North East states, Meghalaya stands third in terms of maternal deaths, and the figures have remained consistent through several years, which is a matter of concern.

Khasi Rural Women and Indigenous Ecological Wisdom

Indigenous ecological wisdom or traditional ecological knowledge (TEK) refers to the knowledge, wisdom, innovations, and practices of the local communities that can be passed down from generation to generation. It tends to be collectively owned and takes the form of stories, songs, folklore, proverbs, and cultural values, beliefs, rituals, customary laws, local languages, and agricultural practices. Sociologists and social scientists in their quest for understanding indigenous knowledge systems have reported how humans have made use of prayer, magic spells, cultivation techniques, land management, cultural beliefs, folksongs, folktales, and so on in their day-to-day activities.

Folktales, folklore, and belief systems have a considerable effect on environmental attitudes and can play a major role in ecological conservation practices. Like any other tribal community, the Khasi are rich in folktales and legends locally known as '*Khanatang bad Puriskam*' which are mostly woven around the various forces of nature such as the hills and vales, rocks and caves, the flora and fauna. These elements of nature are personified in the legends as a mother, son, husband and wife, friend and foe, and so on, each having a significant identity in accordance with their natural behavior and virtues like love, hate, jealousy, pride, and vanity with the sole purpose of teaching moral and spiritual values. These, in turn, instill the elements of nature into the cognitive repertoire and self-concepts of individuals and the identity of the people.

One such legend centers around the so called 'Race between the Two Rivers'. The story goes like this.... It was believed that the God of Shillong '*U Lei Shyllong*' had two beautiful and spirited daughters named *Ka Ngot* and *Ka Iam*. They were twin sisters and were very attached to each other. One day they climbed to the top of the Shillong peak and gazed at the beauty of the landscape below. *Ka Iam*, who was the more confident of the two sisters, challenged a race against her sister *Ka Ngot*. So, they decided the starting point would be the bottom of the Shillong peak and the finish line would be the plains of Sylhet. *Ka Ngot* was a little apprehensive about going ahead with the race, but *Ka Iam* insisted and was overconfident of winning the race. The two sisters transformed themselves into two rivers and set out on the race. *Ka Ngot* was true to her character and quietly and silently meandered her way from the starting point at the bottom of the Shillong peak through smooth valleys and slowly but steadily reached her destination at Sylhet only to find that her sister *Ka Iam* was still far behind. At this, *Ka Ngot* was so dismayed and sad that she panicked and curled her way back and took a sharp turn in search of her sister. Legend has it that this sight was so spectacular and breathtaking, as the silvery waters of *Ka Ngot* was gleaming in the sunlight like a silver necklace, that even today this curved portion of the river is known as '*Rupatylli*' or 'silver necklace'. *Ka Iam* was delayed because being the impetuous one she went through a longer route through the border town of Shella assuming that it was the shorter course but landed on a rocky terrain and lost track of the destination. When *Ka Iam* finally reached her destination, she felt so humiliated on seeing her sister *Ka Ngot* already there that she divided herself into five streams – namely *Ka Umtong, Ka Torasa, Ka Pasbira, Ka Kumarjani*, and *Ka Duwara* – only to finally join the mighty Surma River on its voyage to the sea.

This legend throws light on an important moral value and that is 'Pride goes before a fall', besides connecting rivers to a feminine identity. There are many such legends and folktales among the Khasi that depict and personify natural objects on a gendered principle which simply shows the reverence, spirituality, and importance assigned to natural objects and the nature-woman linkages.

From a gendered perspective it is well known that women's knowledge of ecosystems has contributed to ecological sustainability and is illustrative of the strength of their ecosystem-based learning and commitment to non-violative uses of nature (Krishna, 2004). A cursory glance over the Khasi rural ecosystem reveals the extent of the relationship between women and ecological sustainability. Their wisdom and knowledge about certain species of edible and non-edible plants and herbs, the food culture, and regenerative process of soil fertility has been instrumental towards judicious use of the resources, thereby maintaining a reciprocal relationship between the environment and the people. The resources of nature are utilized only to the extent that they satisfy the needs of the people so that nature can rejuvenate and flourish in time and there is no exploitation.

A study conducted by S. Khongsit (1999), a local environmentalist, revealed that there are about 113 species of edible roots, plants, and herbs that begin with the prefix 'ja' which when translated in Khasi means rice. A common belief among the Khasi is that rice being a staple food, these plants and herbs must have been a supplement for rice at times of crises, such as famine or plague, for which reason they are identified as an important item of food in their daily diet till today. This age-old wisdom and traditional ecological knowledge of the Khasi about certain species of plants and herbs that are of medicinal use has strengthened the bond between women and nature and has played an important role in the spiritual life of the entire family thereby connecting the forest into the social world.

Khasi socio-religious ethics is a complex embodiment in which composition of both flora and fauna as well as cosmic bodies are held sacred. These environmental elements such as animals, birds, and man find space in Khasi myths, legends, fables, and so on and share a close comradeship which has been expressed vividly in these lines: "*U briew U mrad kawei ka ktien ki dang iakren*" (When man and animals spoke the same language).

A field study conducted by the author in Lawbyrwa village of Ri-Bhoi District of Khasi Hills revealed that 80 percent of the time spent by the women of this village in a day was out in the forest, mainly to fulfill the subsistence needs of their family, and these activities included collecting fuel wood, water, edible roots and tubers, broom shrubs, and other non-timber forest products (NTFP). Since most villages are in the forested regions and fringe forest areas, they have properly defined gendered work patterns and responsibilities regarding activities related to their daily chores. While men are actively involved in agriculture, daily wage labour, and farming activities, women are seen to spend more time in the woods and are the main providers of the daily necessities and basic needs such as fuel wood, food, medicinal products, and NTFP to the family.

It is interesting to note the role and importance of nature in their day-to-day cultural and religious life and activities. The naming ceremony of a new-born infant, '*Kajer kathoh*', is an important occasion where they make use of pounded rice flour which is placed in a bamboo winnower called '*Ka prah*'. The use of a bamboo splinter to cut the umbilical cord of an infant is a practice followed till today. There are also many folk dances in which young women take part, such as *Ka Shad Kynthei*, which is one form of dance in which only girls can take part to pray for a good harvest. Another aspect of women's fertility is the *Shad Nohsakyriat*, which is performed by women who spin themselves round and round around a traditional pole to symbolize their strength and femininity as mothers.

However, in the recent past many changes have set in that have not only created a new set of rules and regulations regarding forest management but restriction of accessibility to the forest and forest products.

Forest Jurisdiction

'Land' and 'forest' are gifts of nature to the Khasi. Like any other tribal community, these are the two most important sources of livelihood that permeate every aspect of their life. The Khasi rural folk, especially women, consider the wooded lands as their well-loved home, an abode of worship, and a storehouse of their daily necessities of life such as food, water, fodder, and firewood. Land, which is locally called 'Ri' by the Khasi, has a deep attachment to their pattern of social organization and permeates every aspect of their socio-economic life. Land is 'a gift of nature' that belongs to the community. Therefore, access to land not only ensures economic security for the individual, but control over it symbolizes territorial integrity for the entire community (Nongbri, 2003).

The Khasi rural women venture out into the forest and engage themselves everyday in cutting firewood and collecting fruits, vegetables, broom grass, and so on, drying and processing them, and at the fall of the day, return home with heavy bundles on their head. The hill slopes and deep gorges remain throughout the year covered with wild banana trees, cane and bamboo, and innumerable varieties of edible and non-edible mushrooms.

However, women have largely been marginalized in relation to patterns of power distribution and decision-making, which has hampered their roles as managers in the natural resources management practices. In the matrilineal Khasi society, the issue of equitable distribution of beneficiary in an indigenous management system has often been challenged. Though in principle the matrilineal rule of female inheritance and descent is followed, customary practices are not gender egalitarian, as authority is vested in the mother's brother (maternal uncle). Women can inherit property such as land but lack the power to manage it. The woman is the focal point of the household organization but when it comes to the allocation of rights over land, a sharp distinction is maintained between 'ownership' and 'control'. While ownership of land is transmitted through women, control invariably lies with men (Nongbri, 2003).

Making the plight of women worse is the system of forest management operating in the Khasi Hills. Here the management of forest lands falls under three agencies:

- Autonomous District Council
- State government (Reserved Forests or National Parks)
- Community (*Khlaw Raid* or Community Forests and *Khlaw Adong* or Community Protected Forest)

The diarchic form of control and management of forests in Meghalaya by the State Forest Department and the Autonomous District Councils has contributed to dual forest policies, lack of co-ordination, and division of accountability. Both the authorities formulate their own forest policies and follow a separate set of forest rules and regulations in their respective areas of operation. The main reason for such a diarchy is that while the State Forest Department maintains a well-qualified team of skilled technical staff, it has a comparatively meager area of 1,127.23 sq km of forest under its jurisdiction. The District Council, on the other hand, has large forest covered areas of 6,250.68 sq km under its management and control but it suffers from inadequate technical manpower which has resulted in the lack of enforcement of rules. Forest being the principal source of revenue, the District Council cannot afford to stop forest operation under prevailing fund constraints. The private landowners who own substantial areas of forests indulge freely in the sale of timber from their land turned into forests. In the case of Community Forests (*Khlaw Raid* and *Khlaw-Adong*), greater autonomy in management lies with the traditional institutions, clans, and individuals.

Women's Rights

Land titles are important assets for women because of the growing shift of menfolk over to non-farming activities. Land rights can improve gender relations which would further enhance women empowerment. The plight of Khasi women is further reflected in the indigenous system of forest management prevalent in the Khasi Hills. The diarchic form of control and management of land and forest by the Autonomous District Council and the state government exposes the community-owned lands to exploitation and loss of forest revenue. The ultimate sufferers are the village folk who have lost their accessibility to forest rights and land leading to infringement on their traditional rights over forest and land. It is here that one finds an invisibility of women's role in forest management and decision-making. Although women shoulder heavy economic burdens and responsibilities trying to meet the survival needs of the family, they have little to say in village-level meetings or in the management of their productive resources. Added to this is the problem of infringement of land ownership and the unhealthy trend of gradual shift over from community lands (*Ri-Raid*) to private ownership (*Ri-Kynti*) for developmental purposes or sale of land to private enterprises. The private forest owners who own substantial areas of forests indulged freely in the sale of timber from their forests. According to the 1958 United Khasi and Jaintia Hills Autonomous District (Management and Control of Forest) Act, the private forests in Meghalaya are under the control and supervision of the District Council authorities, but the private owners think themselves free to operate in these forests. A larger quantity of timber than what Meghalaya consumed found its way to the other states in the North East. The Acts and legislation passed by the state for conservation of forests, protection of catchment areas, prohibition of felling and removal of timber, and so on had very little effect until the Supreme Court intervened with a blanket ban on logging timber in December 1996.

It has been observed that the impact of the Supreme Court ban on felling of trees passed on 12 December 1996 has fallen heavily on womenfolk by imposing an additional burden on their daily subsistence needs and increased their domestic chores manifold (Nongbri, 2003). They must walk longer distances to collect fuel, fodder, and forest products for their daily sustenance. Additionally, they must depend upon the male counterparts for sale of produce since the transport facilities in villages are poor and not reliable. This leaves women to be involved only at the lower stages of the value chain such as collection, cleaning, processing, and sorting out the forest produce whereas the major portion of the production chain is being dominated by menfolk who carry out marketing and transport of the produce. This could be the reason for the 'self-perception' by women that they are merely 'helpers' and not 'workers' when it comes to demarcating the flow of work within and outside the household.

The nature-culture dualism that has been nurtured by Khasi women over the centuries has no doubt played an important role in preserving the cultural and religious ethos of the Khasi regarding nature. However, as the environment is facing a crisis today, amidst stringent legislation of the government coupled with problems of development and conflict over the resource base, women's roles as basic providers and managers of a 'green economy' is eroding fast. The practical experiences of Khasi rural women at the grassroots level have basically been experiences that have centered around marginalization due to depletion of natural resources and finding alternative strategies of survival. These along with stringent customary practices have kept women apart from managing these natural resources.

Implications for Conservation of the Environment

Given the dilution of women's rights over land and forests and restrictions on their activities in the forests and control over natural environmental resources there is likely to be/has been a degradation of the natural resources.

Much research has been done to study the impact of new agricultural policies such as the green revolution, forest management laws, and other legislative policies on the trend of rural poverty. Vandana Shiva attacked the role of green revolution technologies in enhancing a destructive rather than constructive developmental paradigm. According to her, environmental degradation is a break from a holistic lifestyle, that is, from people living harmoniously with nature. She is critical of women being devalued and their relationship with nature being severed. Shiva elucidates women's contribution (traditional wisdom) to the food cycle. She claims that importing such western strategies in India destroys women's relationship with food production (Datar, 2011).

Another visible result is the increased poverty in rural areas that has further implicated gender inequities due to underlying economic, social, and historical differences. The concept of 'feminization of poverty' had become renowned around the 90s, which focuses on a gender dimension to poverty. Women living in poverty are often the ones who are denied access to critical resources such as land, forest, and inheritance.

In the North East, predominantly in the tribal regions, the survival economy of basic needs constitutes wood, fuel, fodder, fruits, nuts, green manure, small timber, bamboo shoot, mushroom, broom shrub, medicinal herbs, oils, resins, gum, honey, spices, and more. Most of the rural households gather these essential items from village commons and forests. These items contribute to at least 20 percent or more of the total income of the womenfolk in the villages (Agarwal, 1985). However, studies have revealed that lately due to privatization of land, developmental projects, and government legislations, there has been a serious impact on women in terms of the effort spent in collecting fodder, fuel, and water and the distance traversed in the process. While firewood continues to be the main source of domestic fuel in most villages in the North East, the single most common impact has been on the collection time which has increased manifold due to the longer distance that women must forage daily, which is for over 5 km daily on foot. Scarcity of wood fuel induces the rural folk to consume raw or partially cooked food which affects their health standards, especially maternal and child health.

In the case of the matrilineal Khasi of Meghalaya, the customary rights of women as custodians over land and property is often misjudged as giving them full rights over management of such resources as explained earlier. Women's demands regarding accessibility to resources need to be addressed and articulated separately. Their responses to development strategies have not been studied in a sensitive manner. This gap gives the impression that women have for long sacrificed their traditional means of livelihood for progress and development.

Men and women have distinct roles to play in forestry operations in developing countries. In addition, they have differentiated interests in forest products and services, with women mostly involved in supply of food and energy for the households while men are mainly driven by commercial interests. Thus, any gendered approach to forest management would look at women to be the major stakeholders in forest resources use and management. The traditional wisdom of forest and ethical values linked to forest lore and legends have been part and parcel of Khasi culture. While womenfolk are essential in land utilization, land management and decision-making has been the purview of men and a male-dominated domain even in the matrilineal Khasi society. Even when land is owned by women, it is managed by a man or male relatives. Therefore, it is only their daily activities such as collecting fallen twigs, small firewood, dried leaves and branches, edible roots and herbs that fetch them a scanty source of daily income by selling them in the local markets or along the highway roadside stalls. All this has been happening within sustainable limits. In recent times, due to forest degradation and infringement of land resources, there has been a marked change regarding livelihood patterns of rural forest communities and the worst sufferers have been women. In facing a scarcity of fuelwood, some women have substituted it with other less-efficient fuels such as twigs, leaves, dung, or crop residue. These inferior fuels require more time to light and more effort to keep them burning, thus making cooking more time consuming and preventing women from simultaneously attending to other work (Agarwal, 2001).

Recently, the Scheduled Tribes (Recognition of Forest Rights) Bill 2005 was formulated by the Indian Ministry of Tribal Affairs to affirm the land rights of India's forest-dwelling tribal people. The Bill seeks to recognize rights to forest produce, intellectual property, and traditional knowledge, while conferring duties on the Scheduled Tribes and acknowledging the close connection between the well-being of the forest dwellers and the health of the ecosystem. However, the Forest Rights Bill has been criticized by the Ministry of Environment and Forests, which argues that the Bill will hinder efforts to preserve India's dwindling forest cover and cause irreparable ecological damage. Social groups are also concerned that the Bill will harm the livelihood of forest dwellers who are not considered to be part of Scheduled Tribes. Thus, the Bill in its present form is bogged by controversy with many social and conservation groups and the Ministry of Environment and Forest (MOEF) opposing it. In response to these concerns, the Indian Cabinet has suspended discussion of the Bill until it receives comments from members of the public.

Nonetheless, the drafting of the Forest Rights Bill, which explicitly recognizes historical and social injustice as a serious and legitimate concern, represents a significant step towards the recognition of forest dweller rights in India. The Union Minister for Development of North East Region had in 2006 converted a draft National Tribal Policy with a view to addressing issues like enhancement of human development index of tribals, ensuring their control over natural resources and distribution of wealth and opportunities among the tribals who are hitherto deprived of them.

Delving deeper into these issues it can be hoped for sound legislation that addresses the issues of the tribes in general and the forest-dependent communities in particular. The state government needs to promote a gender sensitive decentralization process that works hand in hand with the local communities and promotes women's economic potential and thereby gives an impetus to the conservation of an environment which is rightfully of the community and their decedents.

Conclusion

The facts and circumstances discussed in this chapter lead us to the conclusion that there is a gradual and inevitable shift in the pattern of Khasi livelihood from a traditional forest-based economy to a more diversified means of living. There is a visible transition in Khasi society regarding the state of their dependence on forest resources like food, fodder, and shelter to other means of living. The depletion of land and natural forest cover, which were their only economic base, resulted in the paucity of forestland for cultivation, affecting womenfolk and their access to forest resources. Pressure of population on existing land has aggravated the situation, leading to poverty and marginalization of women in the economic sector.

The conversion of natural forests and wastelands into Reserved Forest, Protected, and Unclassed Forests by the state and Autonomous District Councils has led to increased restrictions on tribal access to forests, rivers, fisheries, and so on. The plethora of forest laws and regulations enacted by the authorities in the name of development of forest and environment severely infringed the tribal rights to natural forests and forest produces. Women suffer much in terms of accessibility to forest and in many instances, women have been victims of marginalization of resources due to lack of participation in decision-making and land ownership. Government at all levels needs to empower women by engaging them in community-based forest governance. Looking at the gender division of labour in forest-based activities, the government can activate training facilities for women in establishing nurseries of their own, selecting saplings and seedlings, replanting, weeding, and so on, which will provide a household income of their own and at the same time contribute towards sustainable conservation. There is a need to explore the complex linkages and intersections of gender with policy, politics, environmental exploitation, and sustainable resource use globally. There is now available an Environment Gender Index (EGI) that ranks countries on how they are translating gender and environment mandates into national policy and planning. Out of 72 countries that were ranked on the EGI, India is ranked 46. Policy makers have come to appreciate that women 'play an essential role in the management of natural resources, including soil, water, forests and energy and often have a profound traditional and contemporary knowledge of the natural world around them' (World Bank, 1991). While global environmental policies and projects reflect this change, in India a gender perspective still needs to inform policy in meaningful ways. This can be done with restoration of the equitable economic situation that had existed in the past and by respecting women's intimate relationship with nature and their customary rights to land and forest.

Notes

Among the Khasi, the '*kur*' is equivalent to a clan. It is strictly exogamous, and members of the same clan do not marry. The *kur* usually comprises of all the members of the same matriclan, that is, the descendants of a common mother or great-grandmother.

The prefix '*mei*' means mother and '*ramew*' means earth. Thus, *Mei Ramew* is an all-embracing entity comprising of the land, forest, streams, rivers, hills, and everything contained in it. The Khasis uphold her with respect and invoke her blessings to a good crop and thank her for a good harvest.

The maternal uncle is the mother's brother who is designated as *U Kni*. He is the ultimate male authority in his sister's house and has more authority over his sister's house than her husband.

References

Agarwal, A. (2001). *State of India's environment-The second citizen's report*. Centre for Science and Environment.

Agarwal, B. (1985). Women, poverty and agricultural growth in India. *The Journal of Peasant Studies*, *13*(4), 165–220.

Beauvoir, S. D. (1949). *The second sex*. Librairie Gallimard.

Datar, C. (2011). *Ecofeminism revisited*. Rawat Publications.

Gurdon, P. R. T. (1975). *The Khasis*. Cosmo Publications.

Khongsit, S. (1999). Kiba ngi khot *ja*. (Unpublished pamphlet-Upper Shillong).

Krishna, S. (2004). *Livelihood and gender: Equity in community resource management*. Sage Publications.

Levi-Strauss, C. (1967). *Structural Anthropology*. Anchor Books.

Mac Cormack, C., & Strathern, M. (1980). *Nature, culture and gender*. Cambridge University Press.

Nongbri, T. (2003). *Development, ethnicity and gender*. Rawat Publications

Plant, J. (1992). *The green reader: Essays toward a sustainable society*. Dobson, A. (Ed.). Mercury House

Raimi, M., Suleiman, R. M., Odipe, O. E., Tolulope, S. J., Oshatunberu, M., Olalekan, A. S., & Makanjuola, C. B. (2019). Women role in environmental conservation and development in Nigeria. *SSRN Electronic Journal*. https://doi.org/10.2139/ssrn.3425832

Shangpliang, R. M. (2013a). *Socio-economic dimension of forest among the Khasi*. Concept Publishing Company Pvt. Ltd.

Shangpliang, R. M. (2016). Common property resources and rural livelihood among the Khasi of Meghalaya. *Eastern Anthropologist*, *70*, 1–2.

World Bank. (1991). *World development report*. World Bank e Library.

Part III

Conservation Practices

13 Coal Mining versus *Oikophilia*

Changki Village, Nagaland

Amri Tiasunep

Conflict of interests is a maxim in every society. Moreover, no decision taken in the realm of human affairs can be considered as perfect except for consideration of tradeoffs among the available options. Coal mining presents such a situation in the North East region of India, where there is tradeoff between economic reward and adverse environmental effect. The total inferred tertiary coal of North East India, as of 1 April 2021, is 909 million tons (Ministry of Coal, Government of India, 2022a). The quality of this coal deposit is sub-bituminous tertiary (Chabukdhara & Singh, 2015). As India moves increasingly towards an opencast mining method (Ministry of Coal, Government of India, 2022b), so does the trend of mining in the North East. With passage of time and the introduction of heavy machinery, the method of coal mining is evolving from rat-hole mining to opencast mining. While the opencast mining method has its advantages in terms of efficiency of coal extraction, safety of workers and lower cost, the flipside of it is the cost paid in terms of environmental destruction.

Opencast mining contributes to major air pollutants, drastic pollution of water, changes in the landscape, and loss of forest cover (Chabukdhara & Singh, 2015). Moreover, the chemical composition of the coal deposits in North East India is different from the rest of India – high sulfur, volatile matter and vitrinite content which results in more severe environmental impact (Chabukdhara & Singh, 2015).

Nagaland

Nagaland is one of the states in the North East region of India, dominantly occupied by the Nagas, whose cultural heritage varies from that of mainland India (Chophy & Chaudhuri, 2022). In terms of language, the Nagas belongs to Tibeto-Burman speakers with mongoloid features. Geographically, Nagaland is a hilly state where the hills are lower at the foothills of Assam and as one moves further from the foothills, the hills get higher and steeper. According to Nagaland Gazette which is the government communication medium,it is in the foothill belt of Nagaland where coal deposits are usually found (The Nagaland Gazette, 2015). The belt runs through various districts

DOI: 10.4324/9781003504870-16

of Nagaland. Under Mokokchung district and its administration is a village named Changki village; the village and a large portion of its territory comes within the foothill belt. Mokokchung district is dominated by the Ao Naga tribes, and it is to that tribe that the villagers of Changki belong. Though there are many similarities in culture with its tribe members, Changki differs in its culture in some unique traits – their language, political system and art of pottery making (Mills, 2003; Tiasunep, 2021; Tiasunep, 2022). Given the rich coal deposit (Ministry of Coal, Government of India, 2022a), the practice of coal mining was undertaken and practiced at Changki for quite a long period. Individuals associated with the activities of coal mining gained much monetary profit. The desire of such beneficiaries to continue with the activity may not come as a surprise. The beneficiaries include landowners who take royalties for coal mining in their land; owners of heavy machineries such as earth movers, trucks and so on; day wage laborers engaged in manual digging of coal, carrying, and loading of coal into trucks and so forth; miners and those that have invested money for excavation.

While coal mining has been going on for decades, for most of those years, the activity was limited to the task performed within the limits of human muscles, that is, rat-hole mining. Thus, whatever amount of coal a few individuals could carry on their back was the limit of the amount of coal exposed to the environment. Moreover, given the risk involved in the rat-hole mining system, not many individuals were attracted by this enterprise. The advent of heavy machinery changed the picture. Earth movers of various kinds started excavating the land at a rate that could only be dreamt of just a decade back. Those individuals who invested in the use of such machinery were seen to have obtained lucrative profit. Moreover, given that most of the coal is excavated by the machine, the risk involved for one's life has gone down comparatively. Those instant riches, simply based on what the earth can give by opening its belly, attracted many individuals. It was no longer an enterprise of the few. If the use of heavy machinery was responsible for speedy and large-scale excavation, the number of individuals attracted by the enterprise of coal mining has intensified the process of such excavation. Much money has flown into the village because of this enterprise and the miners took initiatives towards monetarily helping the village in certain social causes. There are many blessings to count on one hand. However, there was another hand on which the cost can be counted.

Environmental Cost

The coal mining procedure followed in Nagaland in general is unscientific. It was only recently that Nagaland State Mineral Development Corporation Limited came up with a proposal for scientific mining in Nagaland (Morung Express, 2022). The practice seen in Changki all through these years is no exception. Therefore, ecological and environmental degradation from such practices is expected. A series of papers on scientific studies have shown the ill effects of coal mining practiced in Changki on its environment. Analysis of water quality of the Tzurang river was found to be "degraded and not potable" (Semy & Singh, 2019). Moreover, the coal mine–affected forest at Changki has increased the degree of heavy metal contamination (Semy et al., 2021). Thereby, it resulted in the soil of coal mining affected forest to be "adversely strained from its natural rejuvenating process". In contrast the elements required for healthy soil such as nitrogen, phosphorus, potassium and soil organic matter were depleted in the coal mining affected forest. This impact of mining on the soil properties was extended to 30 cm in depth, which may result in disrupting the root system of the plants (Semy & Singh, 2021a). The impact of mining on the soil has its effect on the plants. Heavy metals such as zinc, cadmium and copper were found to be higher than the standard recommended by World Health Organization in the plant samples of forests affected by coal mining at Changki (Semy & Singh, 2021b).

While confirmation by scientific studies can reaffirm and specify the claims of ill effects on environment, one need not be a scientist to observe such ill effects. Anyone with functional eyes and nose can detect the coal dust at coal dumping sites and it does not take much giftedness in taste ability to detect the sour taste in water making it unsuitable for drinking. To those who were aware of the riverine biodiversity before the advent of heavy machinery for coal mining, the decline in biodiversity is an ever-ready tale to be narrated to any willing listener. One can often hear *"Tzüong ko mesenang sü melia"* which means, "Even a worm does not exist in Tzüong river". The implication of that sentence is, "Forget about fishes, crab and shrimp, we cannot even find worms in Tzüong river".

The villagers are acutely aware of the problems caused by the activities of coal mining based on what they have observed and thereby became the naysayers to this enterprise. The point of adding this is that there are individuals who out of goodwill opined and proposed that perhaps villagers can be educated on the ill effects of coal mining through seminars where scientific findings on the ill effects of unscientific coal mining can be discussed. While such seminars can be of help to a certain extent, as far as the knowledge of ill effects of coal mining is concerned, lay people are not ignoramuses, they know about it, talk about it, and have lived through the consequences.

Even if one does not care about riverine biodiversity, one cannot remain indifferent to the decline in the yield of rice on one's own farm. Coal mining is done at an altitude higher than the Changki valley – the valley where the villagers practise wet cultivation for rice. Therefore, whatever the river brings in its downward flow from such altitude has its impact on the wet cultivation. While not every farmer has felt the impact, those who are recipients of water that flowed from coal mining sites with all its harmful chemicals have seen drastic decline in the fertility of the soil, and therefore, the decline in the yield of rice as narrated by farmers. Some stopped cultivating all together.

Deciding Factors

According to Nagaland Coal Policy (Amendment) (2015), a license for coal mining should be provided through government agencies (The Nagaland Gazette, 2015). However, while applying for such license from the government, "the recommendation of the concerned village council to the local Administration" is required. Thus, to pursue the coal mining enterprise within the jurisdiction of the village, the approval of the village council is required – which is in accordance with the provision of Indian Constitution Article 371A and power assigned by Nagaland Village Council Act in accordance with Article 371A of the constitution. The granting of the approval of the Village Council, however, is not decided by those few individuals seated on the Chairs of the Village Council. Rather, the villagers, those that reside within the village jurisdiction, sit for a general meeting and decide if coal mining is to be allowed within the village territory. If allowed, they decide on the period (in terms of years) for which such activities will be allowed. Such decisions are made with not much consideration of government rules and regulations but on the belief of sovereign ownership of land by the people under customary law.

Enculturation of *Oikophilia*

There are people of Changki origin who reside in other towns and cities; some of them for several generations. Despite living away from the village, they have a strong sense of *oikophilia*, which is "the love and feeling for home". *Oikophilia* is a word coined by Roger Scruton and derived from the Greek words *oikos*, meaning household (from which words such as 'economy' and 'ecology' were derived) and *philia*, which means love (Scruton, 2012). Some have permanently settled outside the village perhaps never to return, so what makes those individuals develop the strong sense of identity and the feeling of *oikophilia* for Changki? The pertinent question remains why the old and the young of those living outside Changki were willing to sacrifice much of their time and resources against the unscientific coal mining being practiced at Changki. Common practice is that parents inculcate *oikophilia* into their children and there are other measures collectively practiced that are further inculcated into the individuals. Since there will be individual variations on how parents inculcate such value to their children, only those measures collectively taken as a society are explored in this present chapter. Every child of Changki descent, implying the father belonging to the Changki origin, after reaching certain standard in high school compulsorily joins the Changki Students' Union known as *Changki Kaketsüri Longjem* locally. Students' Union exists as a unit wherever there are substantial numbers of people from Changki descent. The motto of the Union is "*Tesoyim pongi atemkü*" which is translated as "For the betterment of birthplace". The birthplace here refers to Changki village. Irrespective of where they were born and brought up and whatever their birth certificate may claim, Changki village will remain as their *Tesoyim*, meaning birth village. It is through such a Students' Union that the identity and the love for the village is reinforced among the youth.

As for the adults, there are Changki Unions in various towns and cities across Nagaland, which keep the sense of oneness and belonging intact. In such meetings, the occasion often starts with singing the Changki anthem. One only needs to see the lyrics of the anthem to notice how much it unites the Changki citizens to their ancestral village.

The anthem narrates the history, culture and geography of Changki village. It also has lines expressing the desire to be buried only at Changki, which is one's birthplace. Moreover, every man, wherever he may reside is a member of *Changki Ayim Asem Senso Mungdang* (Changki Citizens' Forum). The meeting of this forum is in Changki village. Those who reside outside the jurisdiction of the village, however, have no say on most of the matters related to what happens within the jurisdiction of the village. They cannot attend such meetings except with the permission of the Village Council.

Another factor that keeps the relationship between the village and the citizens of Changki (citizen of Changki refers to those that reside in various towns and cities outside the jurisdiction of the village) is the land ownership system – most of the land in Changki comes under lineage land-owning system. Thus, even those citizens of Changki who do not reside in Changki and have not made any economic investment at the village and have no intension of ever returning to live in Changki village still have land in the village by virtue of being a member of theirlineage given theypays the membership fees to *Changki Ayim Asem Senso Mungdang*. Thus, such citizens share ownership in matters of land and its resources, known as *Lithi-ongdi*. By the same principle, individuals who may not reside in the village but are citizens of Changki can own private land in the village. It is noteworthy that no outsider (non-Changki) can own land within Changki territory.

Oikophilia – An Asset

Despite all the environmental and agricultural costs, the lucrative profit from coal mining cannot be matched by any other economic enterprise undertaken in the village. This conflict of incentives presents a dilemma. For obvious reasons, there are those that push the idea that coal mining should proceed as usual, while there are those that count the environmental cost and want to shout, stop! For those who are for coal mining, the wet cultivation area located at Changki valley is already destroyed to a large degree, mainly through coal mining, and therefore, since the alternative to closure of coal mining is unproductive agriculture, it makes more economic sense to proceed with the coal mining. It is true that the agricultural practice of rice plantation in wet cultivation areas had been traditional in its practice; it was not that profitable even before the advent of heavy machinery for coal mining.

Moreover, since the practice of wet cultivation started roughly 100 years back, and the boundaries of farms were drawn around that time, almost everyone owned the size of farm that could be managed by human and animal (oxen) power, and therefore, except for some individuals, most of the farmland are of the size that produce rice that is sufficient for a family's annual need. Of course, there is selling and buying of the plots of farmland, thereby increasing the size of the farm, but such selling of farmland is rare. Most sell it only on the condition of dire financial needs or debt. The reluctance to sell such land is because farmland is inherited and selling of inherited land is culturally undesirable because, irrespective of whether the farmland may be profitable or not, that land remains as a memory of the hard work invested by one's predecessors. With all these in context, the profit, in terms of monetary benefit, gained from coal mining far surpasses that of agriculture.

Coal mining activities, however, are not practiced by all. Therefore, the economic outcome is such that some individuals gained a lot out of it whereas the agricultural and environmental cost is borne by all. The fact is that while agricultural and environmental costs are borne by all for the monetary benefit of some individuals, some if not most of the individuals who make such monetary gain are not even Changki people. For example, the owners of heavy machineries, those who invest money for coal mining and the middlemen in the entire chain of coal business. Why the decision is made in favor of coal mining when the whole village sits for a meeting to decide on such matter may be puzzling. Why would those without benefits not raise their voice against it? That question is worth pursuing further.

Coal is not a byproduct of human ingenuity, rather a readymade product of the earth. In the context of Changki, one does not need any expertise on coal mining. If one can accumulate enough capital by tapping on to one's resources, one can always try one's luck by digging the land that belongs to him (or his lineage land) or another person who is willing, provided some royalty is shared. In fact, not much planning is required since the chain of activities flow mechanically; excavate the coal, collect it in a dumping area and sell it to the middlemen. Once the coal is excavated, the customers for purchasing the coal are from outside of Nagaland. Therefore, one needs skills in articulation and negotiation if one must do well in the business. Moreover, since most of the miners don't own the heavy machinery or trucks involved in coal mining, one must hire those machineries. All these activities require making business deals – which requires negotiation skills. The point is, to be successful as a coal miner in the context of Changki, one requires persuasiveness and negotiation skill, provided one is lucky enough to find coal deposits in the site he chooses to excavate – not much technical knowledge is required, as explained earlier. This is precisely the very skill needed to win over the village in favor of coal mining during village general meetings. While in principle every individual can talk and express his desire for or against coal mining, in real practical life, things work differently. Not every individual is gifted as an orator or a persuasive speaker, much less a debater. Therefore, only a few individuals speak in the meeting. This is not to ascribe ill-intention to the speakers; it just works out that way in real life. It so happens that people who are successful in the coal business are those who can speak persuasively in the village general meetings. Therefore, the meeting often concludes in favor of coal mining despite there being many individuals within the village who are not willing to pay the agricultural and environmental costs anymore – especially at the cost of future generations. That is where the asset of *oikophilia* comes to play.

The provision for coal mining within Changki territory was on till the month of August 2021. Around that time, anticipating that provision for coal mining may be extended, the two Baptist churches of Changki village wrote a letter to the Village Council requesting to put an end to coal mining. Perhaps emboldened or motivated by what the churches did, the different units of Changki Union that exist in different towns and cities wrote letters to the Village Council in the same spirit. The letters were the result of several meetings of those citizens of Changki residing in various towns and cities. There were Changki citizens who resided in those towns and cities, who were engaged in coal mining enterprises. However, the majority were in favor of closing it down because they did not want to see their *tesoyim*, their home, being destroyed by treating it merely as an instrument meant for making profit. Perhaps in large measures it was due to that, the village gave some time to the coal miners for *tetem nangiba*, translated as "to tidy up", which is a phrase to mean that miners should clear things up to put an end to coal mining by doing activities such as covering the holes dug out, transferring the coal out from the forest, and so on. At the end of April 2022, coal mining was put to an end in Changki territory. Rejuvenation and healing for environmental and agricultural wounds may take its time, however new infliction has been put to an end.

Behavioral Implications and Change

According to the words stated in Nagaland coal policy 2005, the policy was formulated because of "rampant and unplanned mining activities" in Nagaland. Such mining activities resulted in various types of accidents, health hazards and ecological and environmental degradation. Thus, the Department of Geology and Mining took over the "responsibilities for administration and management of coal mining activities" to explore the available coal resources "through scientific methods of mining" and to "ensure safety of the mines". However, one only needs to read Nagaland Coal Policy (First Amendment) 2014 to infer nothing of such sort happened. The document admits that "unplanned mining activities" are still going on with the usual consequences. In the year 2022, things are none better.

Coal mining stopped in the areas under Changki village jurisdiction because enough people with *oikophilia* raised their voice against it and not because of any state government law. Besides the love of homeland, Changki has the cultural practice of leaving inheritance for descendants which is held with strong conviction; this is a reason why people are reluctant to sell inherited land. The inherited land is not merely seen as a plot of land for economic utility but as a legacy of pride of one's ancestors to be passed on to those yet to be born. The rhetoric of those who wished to stop coal mining was along that line – what kind of inheritance will we leave for those that come after us if we destroy it now? The love of homeland and the strong conviction of wanting to leave good inheritance for the future are due to the inculcation of identity connected to the land. Thus, harming of land, no matter how lucrative the profit-making, was looked upon as harming the very idea of Changki itself. This feeling is not unique to some old people with nostalgic feelings for the village. In the closing ceremony of Changki students' conference held in January 2022 (where students of Changki origin from different units came together for the conference), the students from Kohima (i.e. Changki students from Kohima unit. Kohima is the capital city of Nagaland) enacted a shadow play. The play was well circulated through Whatsapp. The message from those young students was unmistakable – destroying the environment of Changki through coal mining was presented as a "sickness" of Changki itself and with a call to reform from such "sickness".

Whatever is stated about Changki regarding the process of coal mining, the manners of decision making in the village – either for or against coal mining – and the environmental destruction that follows are similar across Nagaland, wherever the deposit of coal resources are found. Empowered by Article 371A of the Indian constitution, people share the feeling of sovereign ownership of land and make the decision accordingly. This is not to say that coal mining is inherently evil. In the context of the lack of proper technical expertise and unscientific mining, and without proper legal establishment or knowledge to undertake externalities to go ahead with coal mining, is nothing less than looting the voiceless future generations and the silent amongst the people who bear and live with the consequence. The cost is also borne by people who do not engage in coal mining.

Thus, in the context of Nagaland, the solution of environmental protection lies not in the laws that state government enact, rather the solution lies in inculcating the love of homeland, *oikophilia*, in the people and let that love work with their feeling of sovereign ownership of land. Many are of the view that providing scientific seminars can be a solution to this problem. But human beings do not live by scientific facts alone. No one will eagerly wait for an earthworm dinner upon hearing a lecture that earthworm has more protein content than chicken or fish. Besides that, except for the technical terms, methodologies and specific measurements, all the people of Nagaland who live near the sites of coal mining know very well of the consequences of coal mining; in fact, they live out the consequences – those enthusiastic about mining know no less as well. And yet, the immediate concrete lucrative profit seems more appealing than the distant, abstract environmental cost for some. Of the two options under consideration, the immediate and concrete lucrative profit of the here and now wins the day, if not for the strong love of homeland that counters such immediate greed. When people have a sense of sovereign ownership of land, environmental protection lies not in the law of the state but in the "habits of the heart" (Tocqueville, 2010). Environmental destruction or preservation are human actions. And human actions are determined by the values of the individual or society.

In the case of the closure of coal mining at Changki, the first initiative was taken up by the Church. Given that Nagaland is a Christian state, environmental protection through the agency of Church is an unexplored territory. While a common stance on coal mining by all the churches in Nagaland is not expected, given that coal deposits are not distributed across Nagaland, coal mining is not the only environmental concern. Varied environmental issues can be dealt with through the Church, thereby inviting the Church as an agent for good. This point stands in sync with the previous point made, that is "habit of heart" is more important for environmental management. The Church with its theology on environmental stewardship can contribute to developing such "habit of heart" that habituates the love of home.

Concluding Remarks

The volume for environmental concerns is becoming louder over these few decades and humanity is still experimenting with various policies on how to best handle it – by searching for solutions ranging from the insights of the simple societies to the marvels of latest technologies. In that spirit of human endeavor for a better and greener tomorrow, it is narrated here how coal mining came to a closure at Changki – a village in North East India – and how its internal culture came to protect the very land that gave rise to that culture. Some of the lessons that can be drawn from this incident are as follows:

- Educating people with scientific facts alone will not be a sufficient approach to protect and preserve our environment. Human beings cannot be reduced to rational machines. Emotions and spirituality are a part of what humans are made up of. The cognitive understanding of the impact of staying close to nature and conserving it deserves equal attention. Thus, the rhetoric of environmental protection measures should target human beings holistically.
- As important as the law enacted by state may be, much more important are the values upon which the law is built, and thus, such values should be inculcated.
- Changki's story teaches us clearly that love is a better motivator of human action towards something positive. Love of home and future generations surely did the job well. Sacrificial acts are often called for in measures for environmental protection and preservation; only love would motivate such acts of sacrifice for collective good. Thus, besides scientific facts, love of our common home should be inculcated.

Reflecting on the voices of concerns about environment from the North East region of India is indeed a bittersweet experience. Addressing these issues is heartening; to a degree it shows that the North East region of India has economically moved out of subsistence economy, because of which it has the luxury to address such issues. However, there are still magnitudes of challenges regarding the economy of North East India. The challenge remains in balancing the pursuit of "our daily bread" and preservation of our home – the planet Earth. To prioritize the former over preservation of our home may seem like an irresistible temptation. The present case draws one's attention to the collectivist nature of our society. Collective measures taken for the greater good with a hint of foresight has led Changki to re-establish its environmental attitude towards a positive curve. Only time can tell if the future generations are prudent enough to understand the natural bond between man and its green environment. However, in the long run, any society that pursues economic gains solely with no regard to its ecology loses both.

Bibliography

Chabukdhara, M., & Singh, O. P. (2015). Coal mining in Northeast India: An overview of environmental issues and treatment approaches. *International Journal of Coal Science & Technology*, *3*(2), 87–96. https://doi.org/10.1007/s40789-016-0126-1

Chophy, K., & Chaudhuri, S. K. (2022). *The cultural heritage of Nagaland*. Routledge.

Mills, J. P. (2003). *The Ao Nagas*. Directorate of Art and Culture, Government of Nagaland. First published in 1926. London: Macmillan and Co. Ltd.

Ministry of Coal, Government of India. (2022a). *Provisional coal statistics 2021–22* (p. 29). Accessed from: http://coalcontroller.gov.in/writereaddata/files/download/provisionalcoalstat/Provisional%20Coal%20Statistics%202021-22.pdf. Accessed 14th November 2022.

Ministry of coal, Government of India. (2022b). *Production of raw coal from underground and open cast mines from last ten years*. Accessed from: https://coal.gov.in/sites/default/files/2021-01/Production-Opencast-and-Under-ground-mines-durin-last-ten-years.pdf. Accessed 14th November 2022.

Scruton, R. (2012). *How to think seriously about the planet*. Oxford University Press.

Semy, K., & Singh, M. R. (2019). Assessment on the water quality of Tsurang River, Nagaland Affected by Coal Mining Drainage. *Indian Journal of Ecology*, *46*(4), 845–849.

Semy, K., & Singh, M. R. (2021a). Comparative assessment on the physico-chemical properties of coal mining affected and non-affected forest soil at Changki, Nagaland. *Indian Journal of Ecology*, *48*(1), 36–42.

Semy, K., & Singh, M. R. (2021b). Assessment of soil physico-chemical properties and heavy metals bioaccumulation on plants at a coal mining affected forest of Changki, Nagaland. *Environment and Ecology*, *39*(1A), 192–199.

Semy, K., Singh, M. R., & Vats, N. (2021). Evaluation of soil quality of a coal mine affected forest at Changki, Nagaland, India. *Journal of Environmental Engineering and Landscape Management*, *29*(4), 381–390. https://doi.org/10.3846/jeelm.2021.15848

The Nagaland Gazette. (2015). Nagaland Coal Policy (First Amendment). *The Nagaland Gazette*, May 15, 2015, Part IIA (pp. 66–103).

Tiasunep. (2021). Revisiting J. P. Mills' Portrayal of Changki Ao Naga. In: Behera, M. C. (Eds.), *Tribe-british relations in India*. Springer. https://doi.org/10.1007/978-981-16-3424-6_7

Tiasunep. (2022). Changki's traditional practice of leopard hunting and its implication on the nature of culture. In G. Kanato Chophy & S. K. Chaudhuri (Eds.), *The cultural heritage of Nagaland*. Routledge.

Tocqueville, A. (2010). *Democracy in America Volume 1*. Liberty Fund, Inc. First published as *De la de´mocratie en Ame´rique* in the year 1835.

Newspaper Source

NSMDC proposes scientific coal mining in Nagaland. *Morung Express*. Date: 30th March 2022. Retrieved from https://morungexpress.com/nsmdc-proposes-scientific-coal-mining-in-nagaland. Accessed: 24th September 2022.

Web Resources

Article 371A of the Indian Constitution:

Downloaded from the Website of Northeastern council, Government of India.

Retrieved from: https://necouncil.gov.in/sites/default/files/uploadfiles/371A.pdf Accessed: 24th September 2022.

Coal resources

Downloaded from the website of Directorate of Geology & Mining, Government of Nagaland. Retrieved from: https://dgm.nagaland.gov.in/coal-resources/. Accessed: 25th September 2022.

Nagaland Coal policy 2005.

Downloaded from the website of Directorate of Geology and Mining, Government of Nagaland. Retrieved from: https://dgm.nagaland.gov.in/wp-content/uploads/2015/04/Coal-Mining-Rules2005.pdf. Accessed: 24th September 2022.

Village Council Act, Nagaland.

Retrieved from: https://www.panchayat.gov.in/documents/20126/0/PR+Act_Nagaland+-+1979.pdf/3255e38f-34be-78e0-531a-6f7e9166667e?t=1554880142311. Accessed: 24th September 2022.

Glossary of Local (Changki) Terms

1. *Tesoyim pongi atemkü*	For the betterment of birthplace.
2. *Tesoyim*	Birth place (In this case Changki vilage)
3. *Changki Kaketsüri Longjem*	Changki Students' Union
4. *Changki Ayim Asem Senso Mungdang*	Changki Citizens' Forum
5. *Lithi-ongdi*	Land inclusive of all its resources.
6. *Tetem nangiba*	To tidy up.

14 Conservation at Selbalgre Wildlife Reserve

Garo Hills, Meghalaya

Lobeni Kikon and Jesmin Begum

Meghalaya, which means "abode of clouds" in Sanskrit, is one of the North Eastern states known as "The Seven Sisters". Meghalaya's geography is uneven in the western and northern sections, with steep slopes to the south and west. It shares a 496-kilometer international border with Bangladesh at its southernmost point. The altitude varies greatly, from 50 metres in the lowest places to the highest peak, Shillong Peak, at 1,950 metres. The state's complex topography explains the wide range of unique flora at various altitudes, as well as the vastly varying climatic conditions and edaphic composition. Meghalaya's forests are typically characterized as temperate. The types of forests in Meghalaya are broadly classified into temperate forests and tropical forests based on the rainfall, altitude, and composition of dominant species. The Khasi Hills, the Janitia Hills, and the Garo Hills constitute the three divisions of Meghalaya's geographical area (Kupar Lyngdoh Tron et al., 2022).

As with other tribal populations in India, the tribes of Meghalaya have an intimate connection with nature, especially with the forests. The tribes traditionally lived close to the forests and depended on the forest resources for all their sustenance. Although development has resulted in much of the population moving to urban habitations, those living in rural areas are still dependent on the land for their daily needs. The inhabitants of Meghalaya are a diverse indigenous group, consisting mostly of Khasi, Garo, and Jaintia tribes, as well as other indigenous groups, albeit in smaller numbers, such as the Hajongs, Kochs, Rabhas, Karbis, and Mikirs. The three major regions, namely, the Khasi Hills, the Jaintia Hills, and the Garo Hills are inhabited by the three main indigenous communities after which each respective region is named. The tribal people of Meghalaya maintain forest areas adjacent to their settlements not only because of cultural duty and religious beliefs, but also for different community needs such as water and other resources available within the forests. Large areas of forest within the Khasi and Jaintia Hills are considered sacred by the local people – these forests are believed to be the dwelling place of their deities. Like places of worship in other religions, spiritually significant places are held sacred, and are therefore restricted from day-to-day human activity. Hunting and gathering, even of seemingly insignificant items such as leaves, wood, and water, is forbidden without the

DOI: 10.4324/9781003504870-17

express knowledge and consent of the community elders, local people, and forest caretakers (Kupar Lyngdoh Tron et al., 2022).

The Balpakram and Nokrek Protected Areas are the two ecological crown jewels of the Garo Hills. The two are connected tenuously by forested lands along the spine of the Garo Hills. The Conservation Reference Series Report documents the declaration of village reserves – the Selbalgre Village Reserve Forest to protect gibbons and the Siju-Aretika Village Reserve Forest and the Rewak Kosigre Village Reserve to secure a critical elephant corridor (Kaul et al., 2010).

Modern communities have chosen to strategize forest management into two seemingly divergent methods – conservation or exploitation. Either a forest may be protected for the conservation of natural habitats and their ecosystems, or a forest may be completely utilized of all the natural resources it may provide for commercial use. Natural resources may be used best in the long term according to principles of sustainable yield, to balance commercial productivity with natural preservation.

Forests have long been an important means of survival for human beings. However, as with all natural sources, forests are a finite source of goods, one which is getting rapidly depleted the world over. With the dawn of understanding this reality has come the initiatives for conservationism. Various conservation programs and checks are now either in place or are being developed, some by governments and others by the local people themselves. One such initiative led to the banning of hunting and killing of the Hoolock gibbon and of forest resource destruction in Selbalgre Village, West Garo Hills, Meghalaya. The people of the North East region are both the implementors as well as the direct beneficiaries of such conservation initiatives. The details of the outcomes of the Selbalgre Village community-led initiative towards forest resource conservation is explored here by interviewing residents of the village and reviewing available literature.

The Hoolock Gibbon

In India, the Hoolock gibbon (*Hoolock hoolock*) is found only in a small part of the North East, south of the Brahmaputra River and east of the Dibang River (Parsons, 1941; Choudhury, 1987). The Hoolock gibbon is found in scattered populations due to the earlier extensive hunting, poaching, and habitat depletion – they are found only in certain protected areas and their long-term survival stands questionable. It is recommended that protective measures should be increased, such as formation of protected areas, protection of existing protected areas, banning of poaching, and discouragement of jhum cultivation in their habitats, through campaigns especially involving churches and village headmen (Choudhury, 2006).

The Hoolock gibbon is a tree-dwelling ape, and is the only ape found in the Indian subcontinent. Its characteristic white brows gave rise to its other name, "white-browed gibbon". The most important feature to differentiate one gibbon species from another is by their territorial song and fur coloration. Some species display polychromatism (varying fur coloration) while some have distinct sex differences (Geissmann, 2014). Some species even undergo multiple colour changes in fur according to various phases of their life. Their body habitus, however, remains much the same.

Mootnick and Groves recognised two distinct species of the Hoolock in 2005, the Western Hoolock gibbon (*Hoolock hoolock*) and the Eastern Hoolock gibbon (*Hoolock leuconedys*). The scientific names of Eastern and Western Hoolock gibbons are considered monotypic but it was formerly considered conspecific with *Hoolock leconedys* (Mootnick and Groves, 2005).

The natural habitat of the Western Hoolock gibbon is found in North Eastern India, in the states of Assam, Arunachal Pradesh, Tripura, Meghalaya, Manipur, Nagaland, and Mizoram. The Eastern Hoolock gibbon, on the other hand, is found only in the states of Arunachal Pradesh and certain areas of Assam (Chetry et al. 2012). Some of the reasons for indirect decrease in Hoolock population are due to the overwhelming habitat loss, jhum cultivation, rapidly expanding tea and coffee estates, urbanization, monoculture tree plantations, human foraging for food, and traditional medicine (Choudhury, 1990, 1991, 1996; Srivastava, 1999; Ahmed 2001; Malone et al., 2002; Solanki & Chutia, 2004; Das et al., 2006; Walker et al., 2007). As the population of this species in the wild has decreased by at least 50% over the past 40 years, they are now listed as 'endangered' in the International Union for Conservation of Nature red list. Hunting and habitat loss are the main factors leading to loss of population.

Choudhury (2006) describes the distribution, habitat, status, and conservation of the Hoolok gibbon in the states of Manipur, Meghalaya, Mizoram, and Nagaland. The Hoolock gibbon still exist in all four states but in much depleted numbers. It has been recorded from altitudes of less than 50 m in Meghalaya to above 2,600 m in Nagaland. The gibbon is nowhere abundant due to hunting, and except for a few protected areas it is found in scattered groups, where their survival is doubtful in the long term. The creation of more protected areas, adequate protection of existing protected areas, control of jhum cultivation and poaching, and awareness and involvement of churches and village headmen in conservation are recommended. Adult males and juveniles of both sexes are black with white eyebrows. When sub-adult, the pelage of the females changes to greyish and then to a tan colour, which is then retained as adults. Across the border, its range extends into a small area of Southern China, Eastern Bangladesh, and Myanmar (Burma).

The tropical rainforests throughout much of South East Asia are the natural habitats for gibbons. As gibbons are unable to swim, large bodies of water such as the Ayeyarwady (Irrawaddy) and the Mekong form natural division lines between different families of gibbons, and therefore, a barrier to distribution of different species genera (Geissmann, 2014). For instance, Eastern and Western Hoolock gibbons are naturally separated by Chindwin River in Myanmar. Chindwin River flows as the common boundary between these two species. Western Hoolock Gibbon is found in North Western Myanmar, that is, West of Chindwin River. The population of Western Hoolock gibbon in the southernmost part of Myanmar was studied by Geissmann et al., who identified and confirmed the presence of Western Hoolock gibbon in Southern Rakhine Yoma, Myanmar, although they were found only in very small numbers (Geissmann et al., 2009).

Factors Influencing Species Composition and Diversity of Forests

There are several variables influencing the species composition and diversity of a forest, including the following.

Fire Frequency

Fire is a major factor affecting species composition and diversity of forests and grasslands. Fires are immensely destructive, and are increasing in frequency in all habitats, including tropical habitats, where fire is used as means of land clearing or land management (Bird & Cali, 1998; Tilman et al., 2000).

Fragmentation of Habitats

Human actions have converted large ecosystems into effectively smaller, separate ones, most often by cultivation, urbanization, road, and power line construction. Population explosion and habitat fragmentation are directly proportional processes that are unlikely to reduce soon, even on a global scale (Tilman et al., 1994).

Modern Transportation and Commerce

Modern transportation has not only brought development to every corner of the world but has had the effect of alien species being introduced into every newly exposed ecosystem, whether by intent or accident. When an exotic species is inserted into a novel biogeographic sphere, there are huge repercussions, such as extinction of existing species, altering habitats, competition for limited resources, introduction of hitherto unknown diseases to which native species have no immunity, and so on. (Lonsdale, 1999).

Community Participation in Forest Conservation

The Government of India introduced a concept through the National Forest Policy of 1988 called the Joint Forest Management (JFM) policy. The JFM policy is a joint venture between the forest departments and local communities in natural forest management. JFM emphasizes the importance of involving village communities in forest conservation, not only to protect existing forests and resources, but to regenerate forest lands adjacent to villages.

Local people now have a heightened awareness and sensitivity regarding the preservation of ecologically sensitive areas. In the Environmental Protection Act of 1986, provision was created as a means of protecting certain areas of land by deeming them as Ecologically Sensitive Areas (ESA). Commercial, industrial, and development activities are immediately banned in any region declared as ESA. Several ESAs in India are of significance regarding conservation reserves and community reserves. These two acts have been instrumental in the protection of natural resources and species diversity.

The continued existence of some species is entirely a result of farmers and other forest dwelling communities' efforts. Local cultural practices and knowledge systems protect and encourage biodiversity. Many tribal communities follow nature worship, which is a religious system founded on the belief that all of nature's creatures are to be protected. This very belief system was responsible for the Sacred Groves of Khasi Hills (the forest of God and Goddesses) existing today, rather unharmed, and unexploited.

The inherently forest-based culture of the people, especially indigenous traditions of nature conservation and forest protection, make further development of community-based protected area networks very feasible. North East India has had community forest management systems in many forms for several centuries – the systems are still the main modality for forest conservation and protection. Community forest protection is the foremost tool in promoting the North East's overwhelming biodiversity. However, it is a fact that the very system responsible for forest preservation has received little attention or support until recently. Previously, community forest protection was not accepted as a salient technique for diversity conservation, despite its proven, widespread, and pre-existent grassroot level network. Organizations have only now begun to appreciate the need to partner with forest-dependent people, including them not only in the implementation of processes, but in policy and decision making.

Existing Efforts

According to Carpenter et al. (2006), forest goods and services are at high risk of being depleted all over the world. Forest lands not only contain important biosystems, but provide food and fuel; in addition, they control floods and erosion and filter air and water. The exploitation of forests, and the subsequent deforestation, has led to a loss of goods and services which is directly due to demographic expansion and use of forests for cultivation, mining, infrastructure projects, and so on. Much of the deforestation around the world has occurred due to population explosion and human demand for resources; a continually increasing trend that will no doubt result in aggravating the already existent ecological crises, which will in turn result in loss of livelihood and quality of life.

The Millennium Ecosystem Assessment (2005) stated that nearly two-thirds of the world's ecosystem services are under threat. Some of the factors for this decline are the local people/communities' lack of awareness of the importance of forest ecosystems, and their inability to understand the necessity for conservation processes. This ignorance and indifference could even result in the failure of governmental initiatives for conservation – it is imperative that the local communities be actively involved in the conservation process, as they will be the very first people to be affected by the loss of forest goods and services. A sustainable model for forest management is of utmost importance as the very ecological systems requiring protection for biodiversity conservation are also significant in the culture and continuance of traditional communities (Kothari et al., 2000).

Community forest management systems have long been the fundamental mechanism of forest conservation and protection in North East India, a technique long unacknowledged by many as a viable means of biodiversity conservation (Poffenberger 2007). This is slowly gaining acceptance, as conservationists recognize the need to harmonize the design of conservation methods with the requirements of ecosystem services and biodiversity (Egoh et al., 2010). Community-based conservation initiatives, are however, usually founded on the understanding of the livelihood-conservation linkages (Brown, 2002; Fernandes, 2005) and utilize participatory activities rather than the approach of advocacy.

Recently, the people's initiatives have become a popular method in conservation and management of natural resources. Often, local communities have been more effective in the rescue of damaged ecosystems in comparison to other methods (Rasika, 2003). Long before the advent of government-initiated conservation projects, community-based forest and wildlife protection were a part and parcel of many communities in the Indian subcontinent. There is documentation of many such conservation practices from Uttarakhand, Assam, Rajasthan, Meghalaya, Manipur, and others. (Kanna et al., 2011).

Hajjar et al. (2011) used the case study approach to ascertain the needs, wants, and current realities of selected forest communities in Brazil and Mexico, to inform bottom-up approaches for community interventions. In identifying community-defined goals it was found that other livelihood strategies, particularly agricultural practices, are needed to be better integrated in the planning of forestry interventions overly focused on timber production. Intervention models on site-specific needs are to consider the variety of contexts and community interests, rather than replicating models that have been successful in other jurisdictions. A better understanding of local perspectives can aid in the design of community forestry interventions brought by conservation and development agencies, by adding an important and under-studied perspective to the problems that face community forestry.

Bowler et al. (2010) has stated that sustainable forest management (SFM) aims to "maintain and enhance the economic, social and environmental values of forests for the benefit of present and future generations". In pursuit of SFM, many developing nations have devolved full or partial forest management authority to local communities. This devolution is expected to result in more effective forest management, conserving biodiversity while contributing to poverty reduction and economic development. Approaches to such community forest management (CFM) have in common the involvement of people who live in and around the forest in the management decisions that affect forest use and conservation. In this context they have defined CFM as 'de-jure' government-approved forms of forest management by local communities, with the following characteristics: 1. a core objective of providing local communities with social and economic benefits whilst promoting the sustainable management of community – or state-owned forests and/or 2. some degree of control and decision-making power vested in the community by the government (or other designated authority).

As a result of imposing a ban on logging and hunting in its resource-rich forest, declaration of a vast area of its forest as tragopan sanctuary, and becoming the first green village in India, Khonoma village has been placed in the national and international tourist map as an important tourist destination and study centre. It has been highlighted that the people of Khonoma are both the beneficiaries and instruments that brought about these initiatives. Interviewing the residents of Khonoma village and reviewing the literatures available, this paper aimed to report the success of the community-led initiative towards conservation of their rich forest resources. The study on Peoples Initiative for Conversation of Forests and Natural Resources found that the responsible people of Khonoma has led the initiatives to ban logging, hunting and destructive methods of resource extraction like setting up forest fire and large-scale collection of forest products. It has been highlighted that this village has been placed on the tourism map of the country and has attracted many wildlife enthusiast, birders, conservationists, researchers, and tourists.

Ruffner and Smith (2019) propose to capture the spirit behind the unique initiative by communities along with Wildlife Trust of India and World Land Trust, which work in partnership with the Garo Hills Autonomous District Council in protecting community lands as 'Village Reserve Forests'. This remarkable way of conservation has resulted in protection of over 2,800 hectares of forest patches and habitat restoration of another 200 hectares. It has been seen that this is an affirmation of indigenous rights, self-government and community empowerment creating a multilevel impact resulting in not just wildlife habitat protection but also a positive social impact on the communities with regards to their livelihood and lifestyle. Such conservation measures can be a big step towards sustainable environmental protection in the long term.

Methodology

The objective of this study is to highlight the intervention and community-led initiative towards the conservation of their rich forest resources. The Selbalgre Forest Reserve is considered a sacred grove and is also a dedicated Hoolock gibbon reserve, which is the pride of the village and community. This reserve is situated around 500 meters away from the village area. The villagers are of Garo ethnicity.

The Garo

Most of the population in the West Garo Hills are the Garos, who constitute one of the major original inhabitants of the district. The arrival of British rule and Christian proselytization brought about conversion of the Garo people to Christianity. Education was the first and foremost reason Christianity spread in the Garo Hills (Miah, 2012). The establishment of schools, clinics, and rice banks by Christian missionaries, as well as the conversion of the Garo to Christianity, have all had a direct impact on their culture (Chicham, 2023). The Garo community is known for its matrilineal society. Sangma (2012) emphasized that the Garos, who reside in the lower Himalayan tracts of the North Eastern corner of India, were traditionally an agrarian community and once followed an indigenous animistic religion. The Garos call themselves "*Achik-Mande*", where *Achik* means hills and *Mande* means people, therefore literally, the hill people. Like numerous other tribes, the Garos have a rich culture, and largely have preserved their cultural traditions. The Garos follow the oral tradition and have a rich repository of folktales, myths and legends, ballads, chants, and recitals. The cultural heritage of the Garos is deeply rooted in their indigenous faith, called Songsarek, a religion devoted to Nature and its manifestations.

It was emphasized by the Nokma of Selbalgre village that in every Garo village, there is an *A'king Nokma*, simply called *Nokma*, or the village head, who is the custodian of the clan lands. The *Nokm*a not only holds much power over village matters, but also has many responsibilities. The majority of the people of Selbalgre Community are Christians. The people in Selbalgre mostly practice jhum cultivation and the native Garo considers the earth as precious and the forest sacred.

Procedure

Data was collected through the qualitative lens with narrative analysis of material (from interviews) comprising of the responses from seven respondents. Participants were recruited through purposive sampling and an open-ended questionnaire was constructed to discuss their concerns. This method was used to gather information with respect to the people's Initiative for Conservation of Selbalgre Wildlife Reserve. However, the amount of information obtained from this method was not sufficient for the study. Due to insufficient data the researcher initiated focus group discussion (FGD) with six community members who had a significant role in managing the forest.

The purpose of FGD was to get further information and to clarify the doubts that the researchers had. The selection of FGD participants at the village level was based on the criterion of the community involved in forest management. Focus group discussion was conducted through open discussion and in a participatory manner. Before starting the FGD, the researcher has given a brief introduction and explained the objectives and scope of FGD. Discussions were initiated by using a set of questions which were translated into local dialect. During the discussion all the participants were given an opportunity to participate and to ask questions.

During the discussions all the FGD techniques were utilized to collect the necessary information which included a brief introduction with resource personnel, team building and rapport building with respondents, and so on. Information on community efforts, Hoolock gibbons, and religious significance were obtained through this FGD with the respondents. Output of the FGD was documented through writing. After presentation of the FGD an open discussion was held to discuss and add or delete the information for finalization of FGD. The discussion was concluded by appreciating the participants for their cooperation and their participation. Secondary literature review related to the data collected has been an integral part of the data.

Analysis

The data was transcribed into written form to conduct a thematic analysis. After reading and familiarizing ourselves with the data we generated an initial list of ideas about what is in the data and what was pertinent or interesting about them.

> This phase involved the production of initial codes from the data. Codes identified a feature of the data that appeared interesting to the analyst and referred to the most basic segment or element of the raw data or information that can be assessed in a meaningful way reading the phenomenon.
>
> (Boyatzis, 1998)

After generating the initial codes, the researcher identified and reviewed the themes which were apparent in the responses. The detailed interview was thematically analyzed.

Results: Themes Extracted

The following are the themes which were extracted from the narratives of the respondents.

Community Efforts

The analyses include the role of traditionally managed forests and the knowledge associated with the conservation of the rich forest resources. Most of the respondents reported that Selbalgre Forest Reserve Committee was formed by the Wildlife Trust of India in 2007 to manage the adjacent forest and to look after the well-being of the communities, as well as the management of natural resources of the areas under their jurisdiction. All activities implemented in Selbalgre are done by the local community and village institutions and the efforts in management and conservation of forests is part of community conservation practices evolved by the Wildlife Trust of India. There is great diversity in the management practices followed by the community. Since the declaration of the Selbalgre Village as a Forest Reserve, the community has been actively participating in all programmes organized by the Selbalgre Forest Reserve Committee. Awareness programmes were also conducted by the Selbalgre Forest Committee and these programmes further ingrained into the villagers the importance of conservation of their unique forest wealth, and thus motivated them to play active roles at all levels of planning and implementation of the conservation initiative. The majority of the activities taken up by the people towards conservation have been funded by Wildlife Trust of India. The Committee manage and control their territory according to local customs and traditions. Their main task is formulating rules and regulations for smooth administration, managing the common property resources (e.g. forest resources). In relation to management of the forests, very clear terms and conditions are laid down regarding the access and use of resources. It was reported that guidelines were framed for all the community members to follow. Some of the rules and regulations that were resolved by the Committee are:

- Preservation of animals like *Huro* (Hoolock gibbons), *Matchapeng* (leopard), *Matchabolga* (cloudy leopard), *Ranggol* (languor), *Matchok* (deer), *Balgitchak* (barking-deer), and *Dogrik-Do-mesal* (jungle fowl).
- Streams and water sources should be preserved; dams must be made and maintained to preserve fish populations.
- Ten hectares are dedicated for saplings of different species to be planted by the community members every year during the months of June and July.

Hoolock Gibbon

Selbalgre Village, which is 32 km from Tura, is home to the unique Gibbon Reserve. This wildlife reserve is protected by the village community and is the ideal habitat for the Hoolock gibbon. As stated by many respondents, the community consider Hoolock gibbons as sacred and traditionally believe that if a gibbon is hunted or killed, a curse would fall on the village. Through the interview it was found that gibbons can be seen only early in the morning. Hoolock gibbons have a religious significance as they have been considered as the custodian of the land and water. They are considered good luck. Although there are no guided tours, locals sometimes volunteer to accompany visitors inside the forest.

Logging and Hunting

With community lives being intimately linked with forest, the people are aware of the direct relation between the quality of the forest and their quality of life. Most of the respondents shared that they understand that large-scale logging operations were endangering the loss of all valuable resources from their forest. The Selbalgre Forest Committee decided to impose a ban on logging in its forest area and the villagers are not permitted to cut down the trees. The Selbalgre Forest Committee issued rules and guidelines as stated earlier and imposed a fine of Rs 5,000; anyone destroying natural resources and hunting animals would be liable for the penalty. Normally, violators of regulations must pay fixed penalties with in-kind services or cash.

Conserving Forest Resources

Even in conservation areas, communities still perceive that forests are a source of both wood and non-timber forest products, despite regulations limiting their use. Community motivations in the study sites are fostered by existing norms in maintaining natural resources for future generations. Most of the respondents cited that community people are encouraged to preserve the Selbalgre Forest to maintain what has always been cherished in the past. A few cited the need to maintain a legacy of the past and to pass on from generation to generation the protection and care for the forest resources. It has been considered as a form of inheritance to be kept for the future generations. They shared that it is a safekeeping to be handed down to the future generations, thus, offering a bridge between the past, present, and future. Respondents cited that preserving such resources is a blessing for the community.

Psychological Implications

Psychological implication refers to any factor impacting or arising in the mind related to the mental and emotional state of a person. The effect of an incident, event, or occurrence on the mental or emotional state of the individuals or groups resulting in a change in perception and or behaviour is included under it.

It has been considered that wildlife hunting is an important economic activity that contributes to the subsistence of indigenous people and to the maintenance of the cultural identity. Changes in the indigenous way of life affect how they manage the resources around them including wildlife population. Hunting is an important subsistence activity for many indigeneous societies and forest dwellers both as a primary source of protein and income source (Luz et al., 2015). Hunting has been a very common practice since ancient times. Past cultures showed that animals they hunted were used for food and clothing, and as time passed by, many tribes eventually mastered the art of agriculture for producing their own food.

It was seen that the community people expressed a deep emotional and psychological connection with the Hoolock gibbon and towards the conservation of forest. They have been very careful to maintain clear rules about logging, hunting, and conservation of forest. The community people of Selbalgre went through a gradual process of realization and as a result they have changed their perception from hunter-gatherer to the protector of the species.

Religious Significance of Hoolock Gibbon

Selbalgre Wildlife Reserve has a cultural and historical value for the Garo community. The concept of conservation has been there for a very long time and the existence of Selbalgre Wildlife Reserve is because it is believed to be the home of 'Guardian Spirit'. Garo ancestors believed in Benevolent Spirits as narrated by one of the respondents. There are Spirits for all kinds of matters both living and non-living things around us. Sacred groves are forest fragments of varying sizes, which are communally protected, and which usually have a significant religious connotation for the protecting community.

For cultivation purposes they always seek permission from the Guardian Spirit. The humans do not disturb the forest areas which are meant for the spirits. The idea of conservation is very spiritual for the Garo communities. This forest (*Asong*) is meant for Guardian Spirit. People must follow certain norms and decorum in their relationship with the forest. It was shared that humans do not disturb the forest. It was shared that collection and removal of any material from the sacred groves is prohibited. Public display of affection cannot be shown, and it is considered taboo. The fear of the Spirit's wrath makes the people conform to the set of rules of the jungle. In the earlier times, if people did not follow what was told, it was considered as a curse and a misfortune. This was considered good for biodiversity. The forest is considered as a sanctuary for wildlife. The fear of the wrath of the Spirit automatically directs the people towards the conservation of the forest.

The Hoolock gibbon has its importance in the creation story of the Garos. They are considered as sacred beings because the gibbon was created to bring life through its call. It resonates throughout the rivers and mountains. It brings life to the physical world, which is why the gibbons are considered as sacred. They are regarded as very intelligent beings.

Psycho-social Relevance Embedded in Community Inclusion

- Maintaining cultural norms regarding forest activities
  Conservation practices and maintaining the cultural norms regrading forest areas were discussed with the respondents during the interview. Through the discussions with the respondents, it has been found that the awareness among the people of Selbalgre regarding conservation and maintaining cultural norms is historical. Maintaining of social and

cultural norms, rules, and practices regarding the forest has helped the community people in keeping the heritage intact for younger generations. Selbalgre people are intricately bound to nature in their social, cultural, economic, ethical, and religious values. For example, one cannot destroy or cut the trees, nor use weapons. The prohibition on selling land represents a cultural norm and the cultural knowledge and practices that support and perpetuate strong relationships between forest use, cultural survival, and cultural identity. It was learned that to keep these sacred groves sanctum sanctorum the people of Selbalgre have not only formulated the rules and regulations but have been passing them to the younger generation.

- Regulation of socio-cultural identity
  It has been observed that traditional practices and beliefs systems that are attached to sacred forest have encouraged local community-driven nature conservation values which have safeguarded conservation of fauna and flora for decades. Sacred forests have been a significant part of many indigenous communities and have been protected by local communities with various cultural beliefs that are unique with special spiritual significance to humans (Dorji et al., 2019). From the interviews it was gathered that indigenous traditional knowledge, beliefs and cultural systems were seen to enhance nature conservation and represent important long-held traditions of conserving specific land areas having cultural and religious significance.
- Continuity in the intimate relationship of the people with the forest.
  The continuity in the intimate relationship of the people with the forest focuses on tree species and their function as a habitat for different species. Continuity is ensured if a minimum growing stock is maintained that provides a forest climate that assures the survival of forest species. Many indigenous communities have a close relationship with nature and animals regarding their religious beliefs. Based on their religious belief the indigenous communities practise rituals which are important for the conservation of nature and wildlife. This can be seen among the people of Selbalgre Wildlife Reserve who have a close affinity towards the natural world, thus acknowledging the intimate relationship between man and nature. According to the community people, life forms are sacred and therefore the act of killing for food or for rituals is forbidden and thus accompanied by the seeking of pardon for their sins (Varah, 2013). Interviews indicated that the people of Selbalgre perceived the biological components of the environment and the human population as an integral part of the natural system. They highlighted that moral norms and values exist which govern human behaviour and that biodiversity conservation is rooted in cultural values, norms, and belief systems. It was further noted that in recent times, these practices and traditions have become of great interest as a source of knowledge for environmental sustainability. It has been found through interactions that the forest has a basic importance and symbolic meanings in the life of the Selbalgre residents. The intimate relationship of the

people with the forest can be traced not only through folktales or myths but is evident from their lifestyles as witnessed.

- Satisfaction of spiritual pining and practices
  The conservation of the Selbalgre Wildlife Reserve has been attributed to spiritual respect and understanding of the practices which include culturally expressed conservation ethics, religious beliefs that conceptualize other species as social beings, and relatively higher biodiversity richness found within the sacred forests in the homeland of these people. Natural resources and people are intrinsically linked. Religious, moral, cultural, political, economic, and ecological boundaries have all shaped environmental use and perceptions. Historically, attitudes and behaviour towards the environment and sustainable use of resources have been greatly affected and determined by nature worship and spiritual values.
- Passing on of rituals, practices, and stories to future generations which are pro-environment
  Forest conservation is the practice of planting and maintaining forested areas for the benefit and sustainability of future generations. Forest conservation involves the upkeep of the natural resources within a forest that are beneficial to both humans and the environment. Forests are vital for human life because they provide a diverse range of resources. The practice of religious rituals and ceremonies by the people allow the forest to be maintained, emphasizing that humans are intrinsically part of the ecosystem. Conservation is the sustainable use and management of natural resources including wildlife, water, air, and earth deposits. Natural resources may be renewable or non-renewable. The conservation of renewable resources like trees involves ensuring that they are not consumed faster than they can be replaced. The conservation of non-renewable resources like fossil fuels involves ensuring that enough are maintained for future generations to utilize. Conservation of natural resources usually focuses on the needs and interests of human beings. The people of Selbalgre believe that passing on the practices, rituals, and stories to future generations will lead towards the growth and development of a better future, but only when the changes take place in ways that are not wasteful.

Conclusion

According to the findings of the study it may be concluded that the people of the Selbalgre have led the initiative to ban logging, hunting of Hoolock gibbons and other animals, and destructive methods of resource extraction like setting forest fires and collection of forest products. This strongly indicates that with active community engagement and with the inputs from Wildlife Trust of India an effective and efficient way of conserving Hoolock gibbons and increasing forest cover has been modelled. The village has been identified as a tourist attraction and has attracted many wildlife enthusiasts, birders, conservationists, and researchers as well. It is seen from the interactions with the community members that they are working together for the conservation of the environment. This gives evidence for the importance of including communities in all programmes, particularly the programmes on the conservation of the environment.

It is crucial to uphold the traditions and beliefs of the community to protect and conserve this unique forest which represents the rich vegetation of the concerned area. Ecological services rendered by sacred groves need to be highlighted and people should be made aware to realize that the conservation of these forest are crucial for their sustenance. Reorientation of local indigenous communities towards restoration of traditional knowledge through environmental awareness programmes by the government or NGOs will be an effective strategy for conservation of natural resources to a desired extent. As is evident it is the awareness and the inherited environmental intimacies of the people of Selbalgre village which have enthused them to take steps and collaborate as a community in the conservation of their forests.

Bibliography

Agnoletti, M., & Santoro, A. (2015). Cultural values and sustainable forest management: the case of Europe. *Journal of Forest Research*, *20*(5), 438–444. https://doi.org/10.1007/s10310-015-0500-7

Ahmed, A. (2001). Illegal trade, and utilization of primates in India. In: A.K. Gupta (Ed.) *Non-human primates of India, ENVIS bulletin: Wildlife and protected area* (Vol 1, pp. 177–184). ENVIS Centre of Himalayan Ecology.

Baral, N. S. R., Bennett, E. M., & Peterson, G. D. (2006). Scenarios for ecosystem services: An overview RID A-9553-2008 RID C-1309-2008. *Ecology and Society*, *11*(1), 29. Retrieved from http://hainanproject.org/wp-content/uploads/2010/04/Scenarios-for-ES-2005-1610.pdf

Bird, M., & J. Cali. 1998. A million-year record of fire in sub-Saharan Africa. *Nature*, 394, 767–769.

Bowler, D. E., Buyung-Ali, L. M., Knight, T. M., & Pullin, A. S. (2010). A systematic review of evidence for the added benefits to health of exposure to natural environments. *BMC Public Health*, *10*. https://doi.org/10.1186/1471-2458-10-456

Boyatzis, R. E. (1998). *Transforming qualitative information: Thematic analysis and code development*. Sage Publications.

Brown, K. 2002. Innovations for conservation and development. *Geographical Journal*, *168*(1), 6–17.

Carpenter, S. R., R. DeFries, T. Dietz, H. A. Mooney, S. Polasky, W. V. Reid, & R. J. Scholes (2006). Millennium ecosystem assessment: Research NEEDS. *Science*, *314*(5797), 257–258. https://doi.org/10.1126/science.1131946

Chase, P., & Singh, K. (2012). People ' s initiative for conservation of forests and natural resources: A success story of khonoma village. *NeBIO*, *3*(September), 61–67.

Chetry, R., D. Chetry, & P. C. Bhattacharjee (2012). Status and conservation of eastern hoolock gibbon Hoolock leuconedys in Assam, India. *Journal of Threatened Taxa*, *4*(13), 3183–3189.

Chicham, S. (2023). Research paper on cultural change of the garo community: A study in Netrokona A monograph submitted as Partial Fulfillment of the Requirement for the Degree of B. S. S. Honors in Sociology. Department of Sociology University of Dhaka. Deadline: 24 No. (March).

Choudhury, A. (2006). The distribution and status of hoolock gibbon, hoolock hoolock, in Manipur, Meghalaya, Mizoram, and Nagaland in Northeast India. *Primate Conservation*, *20*(20), 79–87. https://doi.org/10.1896/0898-6207.20.1.79

Choudhury, A. U. 1987. Notes on the distribution and conservation of Phayre's leaf monkey and hoolock gibbon in India. *Tigerpaper*, *14*(2), 2–6.

Choudhury, A. U. (1990). Population dynamics of Hoolock Gibbons (*Hylobates hoolock*) in Assam, India. *American Journal of Primatology*, *20*, 37–41.

Choudhury, A. U. (1991). Ecology of the Hoolock Gibbon (*Hylobates hoolock*) a lesser ape in the tropical forests of northeastern India. *Journal of Tropical Ecology*, 7, 147–153.

Choudhury, A. U. (1996). A survey of Hoolock Gibbon (*Hylobates hoolock*) in southern Assam, India. *Primate Report*, *44*, 77–85.

Das, J., J. Biswas, P. C. Bhattacharjee, & S. M. Mohnot (2006). First distribution records of the Eastern Hoolock Gibbon (*Hoolock hoolock leuconedys*) from India. *Zoos' Print Journal*, *21*(7): 2316–2320.

Dorji, T., Brookes, J. D., Facelli, J. M., Sears, R. R., Norbu, T., Dorji, K., Chhetri, Y. R., & Baral, H. (2019). Socio-cultural values of ecosystem services from oak forests in the Eastern Himalaya. *Sustainability (Switzerland)*, *11*(8), 2250. https://doi.org/10.3390/su11082250

Egoh, B. E., B. Reyers, J. Carwardine, M. Bode, P. J. O'Farrell, K. A. Wilson, H. P. Possingham, M. Rouget, W. de Lange, D. M. Richardson, & R. M. Cowling (2010) Safeguarding biodiversity and ecosystem services in the Little Karoo, South Africa. *Conservation Biology*, *24*, 1021–1030.

Environmental degradation and human well-being: Report of the millennium ecosystem assessment. (2005). *Population and Development Review*, *31*(2), 389–398. https://doi.org/10.1111/j.1728-4457.2005.00073.x

Fernandes, D. 2005. "More eyes watching": Lessons from the community-based management of a giant fish, Arapaima gigas, in Central Guyana. A Thesis Submitted to the Faculty of Graduate Studies, Natural Resources Institute Clayton H. *Riddell Faculty of Environment, Earth and Resources*, 2(2), 1–169.

Geissmann, T. (2014). *Gibbons-die singenden Menschenaffen: Begleitheft zur Geissmann, T (2014). Gibbons - Die singenden Menschenaffen.* Anthropologisches Institut und Museum der Universität Zürich. Ausstellung.

Geissmann, T., Grindley, M., Momberg, F., Lwin, N., & Moses, S. (2009). Hoolock gibbon and biodiversity survey and training in Southern Rakhine Yoma, Myanmar. *Gibbon Journal*, *5*(5), 7–27.

Hajjar, R., McGrath, D. G., Kozak, R. A., & Innes, J. L. (2011). Framing community forestry challenges with a broader lens: Case studies from the Brazilian Amazon. *Journal of Environmental Management*, *92*, 2159–2169. https://doi.org/10.1016/j.jenvman.2011.03.042

Joint Forest Management. Wikipedia. https://en.wikipedia.org/wiki/Joint_Forest_Management

Kanna, K. S., S. Mishra, & R. K. Kameswara 2011. Creating space for community-based conservation initiatives (CBCLs) in conventional academics. *International Journal of Peace and Development Studies* 2(2), 26–36.

Kaul, R., Tiwari, S. K., Kyarong, S., Dutta, R., & Menon, V. (Eds). (2010). *Canopies and corridors: Conserving the forest of garo hills with elephant gibbons as flagships: Series no. 8.* Retrieved from https://www.wti.org.in/wp-content/uploads/2017/04/pub_canopies_corridors.pdf

Kothari, A., N. Pathak, & F. Vania. (2000). *Where communities care. Community-based wildlife and ecosystem management in South Asia.* Kalpavriksh and International Institute for Environment and Development, Pune.

Kupar Lyngdoh Tron, R., Rangad, D., Shangpliang, W., Manners, B., & Rynjah, I. (2022). The role of community reserved forests in the conservation of anuran amphibians in Meghalaya, North East India. *Protected Area Management - Recent Advances*, 1–25. https://doi.org/10.5772/intechopen.99750

Lonsdale, W. M. (1999). Global patterns of plant invasions and the concept of invasibility. *Ecology*, *80*(5), 1522. https://doi.org/10.2307/176544

Luz, A., Guèze, M., Paneque-Gálvez, J., Pino, J., Maclá, M., Orta-Martínez, M., & Reyes-Garciá, V. (2015). How does cultural change affect indigenous peoples' hunting activity? An empirical study among the Tsimane' in the Bolivian Amazon. *Conservation and Society*, *13*(4), 382–394. https://doi.org/10.4103/0972-4923.179879

Malone, N., A.R. Purnama, M. Wedana, & A. Fuentes (2002). Assessment of the sale of primate at Indonesian bird markets. *Asian Primate* 8(1–2): 7–11.

Miah, S. (2012). Society and culture of the Garo' s in Meghalaya. *IJCAES Special Issue on Basic, Applied and Social Sciences*, II, 376–378.

Mohapatra, P. P., Dash, P. K., Palei, H. S., Debata, S., Sarkar, V., Mishra, A. K., & Dutta, S. K. (2017, December). Ecological and sociocultural aspects of biodiversity conservation in sacred groves of bonai forest division, Odisha, India. In *Biodiversity Conservation and Wildlife Management* (pp. 35–50). North Orissa University Project Protection and Conservation of Sacred Groves of Bonai division in Sundargarh district of Orissa.

Mootnick, A. R., & Groves, C. P. (2005). A new generic name for the Hoolock gibbon (Hylobatidae). *International Journal of Primatology* 26:971–976.

Parsons, R. E. 1941. Rivers as barriers to the distribution of gibbons. *Journal of the Bombay Natural History Society*. *42*: 434 and 926.

Pawar, K. V., & Rothkar, R. V. (2015). Forest conservation & environmental awareness. *Procedia Earth and Planetary Science*, *11*, 212–215. https://doi.org/10.1016/j.proeps.2015.06.027

Plumb, S. T., Nielsen, E. A., & Kim, Y. S. (2012). Challenges of opportunity cost analysis in planning REDD+: A honduran case study of social and cultural values associated with indigenous forest uses. *Forests*, *3*(2), 244–264. https://doi.org/10.3390/f3020244

Poffenberger, M. 2007. Indigenous Forest Stewards of Northeast India. Technical Report. Community Forestry International, Santa Barbara.

Rai, P. K. (2015). A review on Western hoolock Gibbon in Nort-East India. *Journal of Research in Biology*, *4*(3), 1301–1310.

Rasika, D. 2003. Community Conservation Gaining. Report on the transition from conventional conservation practices to community based ones, appeared in India Together News Paper, September 2003. http://www.indiatogether.org/2003/sep/envcommons.htm

Ruffner, J., & Smith, E. B. (2019). *Healing the broken spine: A Community-led conservation initiative in Garo Hills, Meghalaya, India*. Retrieved from http://openarchive.icomos.org/id/eprint/2287

Sarma, U. K., & Barpujari, I. (2011). Eco-cosmologies and biodiversity conservation: Continuity and change among the Karbis of Assam. *International Indigenous Policy Journal*, *2*(4). https://doi.org/10.18584/iipj.2011.2.4.10

Sharma, S., Rikhari, H. C., & Palni, L. M. S. (1999). Conservation of natural resources through religion: A case study from central himalaya. *Society and Natural Resources*, *12*(6), 599–612. https://doi.org/10.1080/089419299279461

Solanki, G.S., & P. Chutia (2004). Ethno Zoological and Socio-cultural aspects of Monpas of Arunachal Pradesh. *Journal of Human Ecology 15*(4): 251–254.

Srivastava, A. (1999). *Primates of Northeast India*. Megadiversity Press, 207 pp.

Tilman, D., May, R. M., Lehman, C. L., & Nowak, M. A. (1994). Habitat destruction and the extinction debt. *Nature, 371*(6492), 65–66. https://doi.org/10.1038/371065a0

Tilman, D., Reich, P., Phillips, H., Menton, M., Patel, A., Vos, E., Peterson, D., & Knops, J. (2000). Fire Suppression and Ecosystem Carbon Storage. *Ecology, 81*(10), 2680–2685. https://doi.org/10.2307/177333

Varah, F. (2013). Situating the humans relationship with nature in the Tangkhul Naga's Lifeworld. *Journal of Human Ecology, 41*(3), 247–254. https://doi.org/10.1080/09709274.2013.11906572

Vatn, A. (2005). Valuing forest ecosystems—An Institutional Perspective. In: S. Kant & R.A. Berry (Eds.), *Institutions, Sustainability, and Natural Resources. Sustainability, Economics, and Natural Resources*, vol 2. Springer. https://doi.org/10.1007/1-4020-3519-5_5

Walker, S., S. Molur, & W.Y. Brockelman (2007). Western Hoolock Gibbon, *Hoolock hoolock* (Harlan, 1831), pp.18. In: *Primates in Peril: The World's 25 Most Endangered Primates 2006–2008*, R. A. Mittermeier, J. Ratsimbazafy, A. B. Rylands, L. Williamson, J. F. Oates, D. Mbora, J. U. Ganzhorn, E. Rodríguez-Luna, E. Palacios, E. W. Heymann, M. C. M. Kierulff, L. Yongcheng, J. Supriatna, C. Roos, S. Walker, & J. M. Aguiar.*Primate Conservation*, 22, 1–40.

15 Community-Based Ecotourism and Conservation of Nature

Manas National Park

Annesha Borah

Humankind's concern for the environment is an important aspect of the development process. Conservation of biodiversity is important for a multitude of reasons, including ethical, emotional, environmental, and economic considerations. Biodiversity is the bedrock of human society, as well as the foundation of economic success and well-being. As a result, its conservation and management are necessary for sustainable development.

Tourism, the most important service industry, is playing a vital role in conservation of biodiversity. Various studies have brought to light that the most popular form of tourism presently is ecotourism, and it has the capacity to provide a powerful incentive for conservation. This broad market segment is increasingly referred to as 'nature tourism,' which contributes to preservation of natural landscapes, flora, and fauna, thereby helping communities in earning livelihood from ecotourism (Wearing and Neil, 1999).

Biodiversity has such high aesthetic value that people travel from all over the world and pay handsomely to explore wilderness and experience the aesthetic value of biodiversity while immersed in local culture. The 'willingness to pay' for the aesthetic worth of wilderness and biodiversity has resulted in the establishment of ecotourism, a sustainable form of tourism based on a monetary estimate of biodiversity's aesthetic value. Ecotourism necessitates a proactive approach that aims to reduce the negative effects of nature tourism while enhancing the good ones (United Nations, 2001). Ecotourism is defined by the International Ecotourism Society as "responsible travel to natural places that conserves the environment and promotes the well-being of local people" (TIES, 2015). This concept not only sheds light on supporting the conservation of natural resources but implies that ecotourism has an essential social dimension. This social dimension encompasses the community-based aspect of tourism. The participation of the local people in ecotourism activities specifically termed as community-based ecotourism (CBET) aids in the generation of revenue, as well as the incentive to protect biodiversity. It serves to generate incentives for local communities to conserve nature by providing alternative revenue sources and livelihoods, as well as empowering local people.

DOI: 10.4324/9781003504870-18

Community-Based Ecotourism

Ecotourism operations that are marked by high environmental concern, increased control, and involvement of local citizens, as well as major benefits for the host community are referred to as "community-based ecotourism" (WWF International, 2001). This is a type of ecotourism in which the local community has significant influence over and engagement in its development and operation, and where most of the benefits stay in the community (Wood, 2002). It is based on the premise that biodiversity must pay for itself by creating economic advantages, especially for local people. CBET's major goal is to combine conservation and local livelihoods, maintaining biodiversity while simultaneously lowering rural poverty, and achieving both goals on a long-term (self-sustaining) basis. CBET is supported by conservation organizations to reduce local risks to biodiversity, such as expanding agriculture, unsustainable gathering of wild flora and animals, and killing wildlife that threatens people's crops, livestock, or lives. CBET acts as a powerful motivator for communities to adopt conservation action. This is such a strong incentive that people will go to great lengths to maintain biodiversity to protect their income (Kiss, 2004). Case studies of CBET projects frequently claim that they were successful in persuading communities to restrict their exploitation of wild plant and animal species, to assist in the management of poaching by outsiders, or to set aside a portion of their farm or grazing land as conservation zones.

Community-Based Ecotourism in India

It has been noted that several case studies in India have demonstrated the effectiveness of CBET in biodiversity protection. A study of CBET in Kerwa Village, Madhya Pradesh, found that after the village's CBET was developed, the local villagers made significant contributions to forest protection. The villagers were involved in tree planting, fencing, checking for illegal tree felling, fire protection, land filling, and sanitation work, among other things. Plantation was done by 70.2 percent of the villagers, fencing by 15.7 percent, illegal tree felling by 8.5 percent, fire protection by 2.8 percent, land filling by 1.4 percent, and sanitation by 1.4 percent (Bhattacharya et al., 2003).

According to a study on the influence of CBET on conservation and the environment, CBET in Andaman has resulted in coral reef protection training and environmental education. CBET provides a cost-effective substitute for fishing, hunting, and harvesting forest and mangrove products (Goodwin & Rosa, 2009).

In North East India, moderate number of studies have been done on CBET. North East India being a perfect blend of nature, culture, and modernity, community-based ecotourism is gradually developing in various parts of the region. This has attracted the attention of researchers to make in-depth studies on the topic. Hussain (2021) has highlighted the positive impact of community-based tourism in Kaziranga National Park and has brought to light that community participation in tourism has led to preservation of the local culture alongside the opportunity to showcase the local cultural and natural assets to the outside world. The author calls it an exposure of socio-cultural assets to the world. The author has inferred through in-depth study that due to local community participation ecotourism has not only enhanced the level of income of the stakeholders in Kaziranga but has also encouraged the locals to conserve the natural resources.

When modernity is everyone's way of life, it is a challenge for a community to preserve its unique folk culture and craftsmanship that exist among its people. But through community-based ecotourism local communities can discover the importance and commercial demand of their resources which serves as a motivation for preservation of the same. Borah and Ryngnga (2019) highlight that community-based ecotourism is a key strategy for reassessing and commercializing indigenous and regional art and culture.

Forest and Nature Conservation in Assam

Assam occupies a total area of 78,438 sq km and accounts for nearly 2.4 percent of the total geographical area of India. The state of Assam is geographically characterized by fertile valleys, dense forests, numerous perennial rivers and streams, lofty hills, the Karbi plateau, and undulating plains. The state enjoys wet and humid summers and dry and cool winters. The climatic condition and wide variety in physical features have resulted in a diversity of ecological habitats such as forests, grasslands, and wetlands, which sustain a wide variety of flora and fauna.

Assam is one of the richest biodiversity zones in the world. This is because Assam falls in the transitional zone of Indian, Indo-Malayan, and Indo-Chinese bio-geographical regions. Assam is also a constituent unit of the Eastern Himalaya Biodiversity Region. The recorded forest area of Assam is 26,832 sq km accounting for 34.21 percent of its total geographical area. The natural vegetation or the forest cover of Assam forms the backbone of its culture, society, and economy. The culture of Assam and its festivals are closely related to nature. The economy of Assam is mainly based on agriculture and forest-based products. The social norms, customs, and traditions reflect the very close and harmonious relationship of man and nature in the state. Every tribe and community has their spiritual and traditional practices that contribute to the conservation and protection of nature. The socio-cultural ethos and practices reflect the man and nature relationship in the state.

However, over the course of time and with the rise of anti-social practices, the state of Assam experienced decline in forest cover. During the 1980s and 1990s, Assam saw large-scale loss in forest resources due to the rise of militancy and sudden change in the demography of the state leading to illegal encroachments. The deforestation activities and poaching happened in the National Parks of Nameri, Kaziranga, Manas, Dibru-Saikhowa, and all other

Table 15.1 Forest cover of Assam

		2017		*2019*		*2021*	
Sl no.	*Category of forest cover*	*Area (sq km)*	*Percentage of geo. Area of State*	*Area (sq km)*	*Percentage of geo. Area of State*	*Area (sq km)*	*Percentage of geo. Area of State*
1	Very dense forests	2797	3.565	2,794.86	3.56	3017	3.84
2	Moderately dense forests	10192	12.99	10,278.91	13.10	9991	12.73
3	Open forests	15116	19.27	15,252.74	19.45	15304	19.51
5	Tree Cover	1496	1.91	1,408	1.80	1630	2.08
6	Total Forest & Tree Cover	29601	**37.74**	29,734.51	**37.91**	29942	**38.172**

Source: India State of Forest Report 2017, 2019 and 2021.

forest covers. However, of late, there was a realization among the local youth to protect and save their environment. The local communities through joint forest management with the state government has taken significant steps for afforestation and conservation activities. At present 34.21 percent of the geographical area of the state of Assam is covered by forest area. The protected area of the state includes seven national parks, four tiger reserves, five elephant reserves, two biosphere reserves, 17 wildlife sanctuaries, one Ramsar site, and two World Heritage sites.

Humankind's concern for the environment is gradually rising all over the world and in Assam as well. It is interesting to note that as per the India State of Forest Report, forest cover in Assam is gradually increasing which is evident from the India State of Forest Report 2017, 2019, and 2021. The increase in forest and tree cover in Assam can be attributed to the plantation activities mainly outside the forest areas. This indicates local people's awareness of environmental conservation. The figures from the India State of Forest Report 2017, 2019, and 2021 are shared in Table 15.1 below:

The Bodo Tribe of Assam

The Bodo are one of the largest Scheduled Tribes in North East India and one of the major indigenous communities of Assam. Historically, they have been recognized under various names such as Kiratas, Mache (Mech), Kocharis, Bodo, or Boro, depending on the region. Linguistically, the Bodo belong to the Tibeto-Burman sublanguage family and have inhabited the North Eastern region under the broader Mongolian ancestry (Britannica, 2017). The majority of the Bodos primarily reside on the northern bank of the Brahmaputra Valley; however, their presence extends across Assam, North Bengal, and certain regions of Arunachal Pradesh and Nagaland. The Bodo community boasts a rich traditional cultural heritage that plays a vital role in maintaining a harmonious ecological balance within their society. This holistic cultural approach is evident in various aspects of Bodo life, including their names, social institutions, feasts, festivals, symbols, rites, and rituals. The socio-religious life of the Bodos reflects their conception of creation and its interconnectedness with both humanity and the natural environment. *Bathou*, the supreme deity in Bodo animistic beliefs, is worshipped through rituals and ceremonies known as *Bathou* Puja. These rituals involve expressing gratitude for the creation of the natural world, emphasizing the interconnectedness between the divine, humanity, and the environment. Bodo festivals like *Bwisagu* and *Domashi* involve rituals that reflect the community's connection with the cycles of nature, expressing thanksgiving for the harvest and seeking blessings for a prosperous future. Animistic beliefs shape the Bodos' worldview, considering various elements of nature, such as rivers, hills, and trees, as sacred. This belief system underscores the idea that these natural elements are integral parts of the creation process (Bhattacharjee & Boro, 2017).

In terms of economic activities, agriculture holds a central position, and the Bodo depend entirely on the providence of Mother Earth and its creator for their sustenance. Bodo festivals often incorporate elements inspired by the natural surroundings, reflecting a deep connection with the environment. One instance is *Bwisagu*, the Bodo New Year festival, when celebrations include traditional dances and rituals expressing gratitude for a bountiful harvest, showcasing the integration of nature into cultural festivities (Mochahary, 2019).

Bodo folklore and mythology are rich with stories that highlight the community's deep connection with nature. These stories often emphasize the importance of living in harmony with the environment. Their folklore depicts animals as spiritual entities, reinforcing the Bodos' respect for the natural world. The artistic expressions of the Bodos, including traditional arts and crafts, are infused with elements inspired by nature. Craftsmanship often involves the use of locally sourced materials, showcasing a sustainable approach to creativity. For example, Bodo baskets, made from locally available bamboo and cane, showcase intricate designs inspired by the environment. Bodos engage in various handicrafts, traditional jewelry making, and pottery. Traditional Bodo textiles often incorporate motifs representing natural elements like flowers, rivers, and animals (Mochahary, 2019).

At present times, the Bodo community is facing environmental challenges because of deforestation, industrialization, urbanization, and climate change. These threats pose a risk to the delicate environmental balance that the Bodos have maintained for centuries. Despite these challenges, the Bodos exhibit resilience in preserving their traditional environmental practices. Efforts are made to balance tradition with adaptation to contemporary challenges.

The Bodo community actively engages in community-based conservation projects today. Collaborative efforts with governmental and non-governmental organizations aim to preserve their ecological heritage and promote sustainable practices.

The Present Study

It is in this context of growing environmental threats that the current study is undertaken to explore the role of Bodo community in ecosystem conservation in and around Manas National Park. The study is mainly based on the local community's initiative of ecotourism with an aim of restoring the once damaged biodiversity of the Manas National Park in the North Eastern state of Assam. The park, which is located on the North West bank of the Brahmaputra River, is about 175 kilometers from Guwahati, Assam's capital city. It is a World Heritage Site, a national park, and a tiger reserve.

Objectives of the Present Study

- To study the growth and development of community-based ecotourism activities in Manas National Park.
- To understand the effect of community-based ecotourism in restoration of biodiversity in Manas National Park.

Methodology

Study area: Manas National Park and Tiger Reserve

The study is concentrated in Manas, or Manas National Park, which is in the Himalayan foothills of Western Assam. Manas was first designated as a game reserve in 1928. It was then upgraded to a tiger reserve in 1974, a World Heritage Site in 1985, biosphere reserve in 1989, and national park in 1990. The park, which is 500 sq km in size and contains large areas of grasslands, is well known for its distinctive scenic splendor. More than 20 endangered species are residents of the park. The golden langur, one of the rarest simian species in the world, was spotted for the first time in Manas in the middle of the 20th century. Rhinos, elephants, tigers, pygmy hogs, hispid hares, Assamese macaques, rhesus macaques, leopards, swamp deer, wild pigs, and other species are also frequently seen in this area. The ideal time to visit the park is November to April.

According to the Assam Forest Department, the park has 61 periphery villages. Agriculture and other activities related to forest resources are the main sources of income for these people. Apart from that, they are engaged in tourism activities which have helped in developing their quality of life. At present the Manas National Park stretches from Kokrajhar to Udalguri district in Assam.

Manas National Park is one of the most important natural world heritage sites in India. It is situated at the confluence of the Indian, Ethiopian, and Indo-Chinese realms, resulting in a rich biodiversity. However, during the 1980s the ecosystem of Manas had been greatly disturbed by massive civil unrest. Insurgency (1988 to 2003) caused huge depletion of forest habitat and wildlife population in Manas. The Bodo movement of the 1980s had a significant impact on the national park, first because insurgent organizations and militants exploited the forests as hideouts, and second because both national and international poaching groups took advantage of the situation, resulting in widespread destruction. Due to serious damage to the forest ecosystem of Manas, UNESCO in the year 1992 declared it as a 'World Heritage Site in Danger'. This situation arose mainly because of two reasons: heavy poaching and terrorist activities. Later, the Bodo Accord in 2003 resulted in the formation of the Bodoland Territorial Council (BTC), which was the turning point for the World Heritage Site as well. This created an awareness among the people that Manas, once the pride of the Bodo, needs to be restored. Later the Forest Department and local youth relentlessly worked for the restoration of the glory of Manas National Park and Tiger Reserve.

Due to intense combined efforts of the government and local communities the wildlife population of the park gradually increased, and the forest cover was restored. On 21 June 2011, UNESCO removed Manas from the List of World Heritage in Danger and those associated with the environmental conservation work received huge appreciation for relentless efforts in preservation in a short period of time.

Research Design

The study is exploratory and descriptive in nature. It puts forward an extensive description of how local communities have put effort on community-based nature conservation through ecotourism. To achieve the objectives of the current study, mixed methodology has been applied. Integration of qualitative and quantitative methodologies has been done to understand the development of community-based ecotourism in the area and its effect on nature conservation through primary as well as secondary data.

Sampling and Data Collection Techniques

Secondary data was collected from journals, books, and newspaper publications in the context of the study. Collection of primary data involved an interaction with the predefined respondent categories. The primary data has been received by interacting with people from nearby communities, who are directly involved in conservation practices and are associated with Manas Maozigendri Ecotourism Society. Manas Maozigendri Ecotourism Society is a non-government organization registered under the Societies Act 1860. It gets its name from the Maozigendri, a river that runs through Manas National

Park. The word 'Maozigendri' is a word used to address little girls out of love and affection by elderly ones. The word Maozi in Bodo means 'cat' and 'Gendri' means a girl with a pot belly. The Society has been working to protect the biodiversity of Manas National Park and development of the socio-economic aspect of the local people. The organization has been operating through community-based ecotourism model in the Eastern region of Manas National Park.

The community involved with Manas Maozigendri Ecotourism Society are mostly the Bodos who share an intrinsic relationship with nature. As mentioned earlier, Bodos' environmental practices are deeply embedded in their indigenous roots, reflecting centuries-old traditions of sustainable living and ecological harmony. Bodos possess traditional ecological knowledge passed down orally through generations. This knowledge encompasses insights into local flora, fauna, weather patterns, and sustainable agricultural practices. Bodo society is community-centric, and decisions are often made collectively emphasizing a sense of unity. Therefore, despite external pressures, the Bodos exhibit resilience in preserving their traditional environmental practices through a community-centric approach. The community actively engages in finding solutions to balance tradition with adaptation to contemporary challenges.

In this study focus group discussion was conducted to derive the required information regarding the community conservation practices with special reference to their community-based tourism initiatives. Focus group discussions (FGDs) are moderated conversations with a small number of people who are knowledgeable or interested in a specific subject. FGDs are frequently conducted with six to 15 participants. They often interact on a specific set of guided questions meant to elicit detailed information. The average length of a discussion is usually one to two hours. FGDs are frequently used to get the opinions of those who are the required sample as decided by the researcher. Focus group discussion is often used in qualitative research to acquire in-depth understanding of social issues under the study.

For the present study, the conservation practitioners of the Koklabari area were assembled for participation in the focus group discussion. The FGD comprised of three groups of 12 members each. Therefore, a total of 36 people from the Koklabari area associated with conservation activities participated in the interaction. The questions asked to the participants were open ended, which provided them with ample scope to place their views regarding their conservation activities. The local community narrated their conservation efforts and achievements till date. They described their involvement in social work during nationwide lockdown when tourism activities worldwide were at a low.

The study used analysis of narratives of the local communities regarding the history of reviving the biodiversity of Manas through their efforts, tourism activities, and the steps taken by them to reach their desired goals of economic returns through ecotourism development as well as conservation practices under the purview of Manas Maozigendri Ecotourism Society. These groups of people expressed their views regarding their efforts and dedication for development of the ecosystem in Manas during the FGD which was recorded. As mentioned previously, the FGD was confined to the people who are directly involved in conservation activities and associated with Manas Maozigendri Ecotourism Society. Narratives derived for the study were analyzed from the information recorded in the FGD. As a part of narrative analysis, a case study has been presented in the study to highlight the people's involvement in conservation activities and the reason behind their motivation.

Discussions and Analysis

As the socio-political turmoil in Bodo area subsided after the Bodo Accord, there grew a consciousness among the local youths for economic development of the villages in and around Manas and the restoration of the biodiversity of Manas National Park. The environmentally conscious youth of these villages formed organizations and joined hands with the Forest Department in joint patrolling in Manas for protection of the forest area and wildlife. The leaders of the organization encouraged local villagers to promote community-based tourism. It was realized that involvement of local people in ecotourism activities would help them grasp the worth of resources, resulting in greater resource protection, strong conservation, sustainable touristic use of resources, and enhancement of quality of lives. These organizations formed by the youth of the villages in and around Manas National Park and Tiger Reserve have been doing commendable work for conservation, preservation, and protection of Manas. The names of these organizations are listed as follows:

- Manas Maozigendri Eco-Tourism Society
- Manas Agrang Eco-Tourism Society
- Manas Bhuyanpara Eco-Tourism Society
- Swmkhwr Mithinga Onsai Afat
- MNP Protection Panbari Eco-Tourism Society
- Kahitema United Social Welfare Society

The current study focusses on Manas Maozigendri Ecotourism Society (MMES), which is a successful example of best practices in ecotourism and environmental conservation. Manas Maozigendri Ecotourism Society (MMES), formed by the efforts of local leaders and responsible youth in Koklabari Mouza, is in Baksa district. It is a non-governmental organization (NGO) dedicated to conservation activities and ecotourism. It is also committed towards the overall development of the fringe village communities of Manas National Park in Bodoland Territorial Council (Assam) area and the ecosystem of the forest area in the region. It was founded on 13 December

2003. On 3 October 2005, the Society was registered under the Societies Registration Act 1860, with Registration No. RS/BAR/237/F/55 of 2005–2006. Community-based tourism initiatives in Manas National Park were undertaken by the Manas Maozigendri Ecotourism Society (MMES) which has placed Manas in the worldwide fame of tourist destinations today.

Manas Maozigendri ecotourism society has been doing a commendable job in creating awareness among the local people. Poachers were converted into conservationists through a community-based movement. The origin of the ecotourism society can be traced back to the Bodo movement in the area. The violent Bodo movement of the 1990s led to large-scale deforestation in the forests of Manas and a great loss to faunal diversity. The formation of the militant group Bodo Liberation Tigers and their anti-government movements was the gloomy epoch in which the natural resources of Manas were depleted. There was extensive deforestation at Manas' southern frontier, which stretches over the entire Bodo-dominated territory. Due to the complete breakdown of law and order in and around Manas, residents frequently sought refuge in the forest to avoid policing forces' atrocities. Taking advantage of the lawlessness, some unscrupulous merchants that had free access to the area formed organized poaching and cutting of valuable trees. Poachers attacked the forest offices on several occasions, destroying most of the park's protection camps. The forest guards were forced to evacuate practically all their protection camps due to the catastrophic situation. The situation was so grave that UNESCO designated it as a "world heritage site in danger."

The step to restore the biodiversity of Manas National Park and develop it as an international tourist destination was incorporated in the Bodo Accord for creating Bodoland Territorial Areas Districts in 2003, when it was finally signed by the Bodo Liberation Tigers and the Central Government. The Chapaguri Kaklabari Anchalik Committee of the All-Bodo Students Union was inspired by this development and founded the Manas Maozigendri Ecotourism Society on 13 December 2003, which was registered under the Societies Act 1860. They held motivational campaigns in regions where tree felling and poaching were common, urging local communities to abandon their damaging practices and seek alternative sources of income. The main goals of MMES are:

- To conserve Manas Tiger Reserve with special preference to its core zone and to restore the past glory of Manas and make it free from the endanger tag of UNESCO World Heritage Site.
- To promote Manas National Park into an international ecotourism spot.
- To establish Manas Maozigendri conservation and ecotourism model.
- To preserve the most ideal traditions and cultures of Bodos.
- To motivate and create awareness towards conservation of nature and environment amongst the people.
- To bring about socio-economic and educational development in the fringe villages of MNP through ecotourism and community tourism.

- To promote Manas Maozigendri Ecotourism Concept across the globe.
- To develop network and co-operation among the likeminded NGOs throughout the world.
- To rehabilitate the ex-poachers.

Key Findings of the Study

To understand the community's role in biodiversity restoration in Manas National Park the responses in the FGD were analyzed and the following activities are noted that have been taken up by MMES.

- Training and awareness programs
  MMES holds programs to educate communities living in and around the park about the importance of Manas conservation and awareness campaigns on conservation of nature and importance of ecotourism for the sustainable livelihood development of fringe villagers. Eminent speakers are invited on the special occasions of World Environment Day, World Tourism Day, the birth anniversary of MMES, and so on to enlighten the local villagers on the importance of biodiversity management. MMES collaborates with many national organizations like All Bodo Students' Organization, Bodo Sahitya Sabha, Assam Science Society, and others for the conservation of flora and fauna. Recently the organization planted around ten thousand trees at Daodhara Reserve Forest adjacent to Manas National Park. The area was earlier encroached upon for inhabitation and farming. MMES has been successful in capturing poachers, educating them, and then hiring them as volunteers with pay and benefits. Training and workshops are conducted from time to time by MMES.
- Rescue operations
  MMES workers are proactive in rescue operation of wildlife from nearby areas. As soon as they are informed regarding wild animals being seen in the villages, they rush to the spot and rescue it. It is freed back inside the core area of the Manas National Park after treatment.
- Welfare programmes
  Sometimes people become homeless due to destruction by wild elephants in the fringe villages. MMES helps these people in the reconstruction of their shelters. Further, MMES helps the poor families with their education, medical, and cultivation requirements. MMES runs several self-help groups with members specialized in cultural qualities, handlooms, handicrafts, and so on and regularly conducts training programs for them.

  During the nationwide lockdown of 2020 due to the COVID pandemic MMES started a campaign to help the needy people of fringe villages of Manas National Park. The campaign was able to collect a total amount of Rs 300,000. Additionally, team MMES received material support from Bhutan India Friendship Association, Nganglam division, for 342 households. MMES received aid for 151 households from WWF India. Using the money raised and the material support from these organizations they were able to reach a total of 2140 households in 24 fringe villages of Manas National Park. During this entire process MMES worked closely with the government officials and were sensitive to individual family needs based on their size, supplies received from the ration shops, and their remoteness. Though the government supplied rice grains and food items, it was only for Ration Card holders. Therefore, MMES took care of the households without Ration Cards till lockdown was over.

- Ecotourism practices
  Ecotourism and conservation are two sides of the same coin. To strengthen the conservation effort, the ecotourism industry creates livelihoods for stakeholders and conservation initiatives enhance the ecosystem resources, which helps the ecotourism industry to pull more tourists to the area. MMES has worked enormously to promote ecotourism activities in Manas. Various tourism activities offered for tourists in Manas are:

i Birding and wildlife sighting
In the field of birding, Mr. Rustam Basumatary, General Secretary of MMES, belonging to the local Bodo community, has been working for the identification of birds along with other faunal species of Manas National Park. He has been able to identify more than 500 species of birds found in Manas till now along with their most favoured habitat in the wild. Eminent bird lovers and watchers from India and around the globe, come to watching birds and demand Rustom's guidance by accompanying them. In 2005, when bird watchers Mr. Bikrom Grewal, Sumit Sen, and Mr. Sujan Chaterjee came to Manas to observe birds of Manas, they were guided by Mr. Rustom Basumatary. They were incredibly impressed with Rustom's guidance and bird identification skills. Later they surprisingly reported Rustom as the "The Eagle Eye" for his excellent identification and guiding on the birds observed in Manas National Park. Mr. Rustom Basumatary has been recently awarded the prestigious Assam Gaurab Award by the government of Assam for his selfless dedication in the field of conservation and ecotourism.

ii Jungle camps
Manas Maozigendri jungle camp, located at the Eastern border of the park, has been constructed for stay of tourists and provides one of the most authentic wilderness experiences. Eco-camps and the tourist activities are run by community members with a benefit-sharing system in place (75 percent goes to individuals providing services, 20 percent towards a community fund, and 5 percent to the park). In 2005, MMES realized a long-held ambition to turn Manas into an international tourist destination when the first foreign visitor stayed at their jungle camp.

iii Jungle safaris
Open jeep safaris, jungle trails, rafting on the Manas, and cultural tours are some of the activities provided to guests by the team of MMES. Tourism has been contributing much-needed funding for their conservation operations, as they have extremely limited sources of revenue.

Case Study of a Conservation Worker

During the FGD, a conservation worker came forward to express his experience and transformation as a conservationist. He was Shri Buddhiswar Bodo, aged 45. He was previously known as a poacher who was involved in wildlife hunting and exporting rhinoceros' horn, elephant's ivory, and skins of animals to other nearby states. He expressed that the sole reason for his involvement in the illegal activities was due to the poverty-stricken financial condition of his family. The impoverished condition and the responsibility of feeding his family led him to earn his living by trading the animal products for which he was paid highly. However, one day while he was poaching, he was attacked by a tiger. He managed to survive but he lost one of his arms. After the incident, the workers of MMES approached and convinced him to surrender himself to the government and dedicate himself to the conservation of Manas. In 2004, he surrendered his personal rifles to the government and since then he has been working as a member of MMES actively for environmental protection. He was awarded recognition by the state authorities. He has received various financial assistance from the government under schemes like Pradhan Mantri Awaas Yojna-Gramin (2021–2022). He expressed his satisfaction for being able to work for the good of society and nature. He now motivates upcoming generations to work for Mother Nature and the holistic good of the land and people.

Conservation Activities of the MMES: Progress through Challenges

Due to a lack of funds and intense opposition from the local community, the first two years of MMES were very challenging. To begin, awareness campaigns against poaching and deforestation were held in a few of the outlying villages. Members of the Society individually went to the homes of hardened poachers and explained the flaws of the trade to them. As a result of the social pressure, a group of 50 poachers gave up their arms to work as conservation volunteers with the Society. The conservation volunteers conducted a study in the forests with the cooperation of these poachers to identify the regions where the most poaching and logging occurred. After witnessing the degree of passion and commitment, the forest department and the society joined together to gradually remove all poaching camps from the park through JFM initiatives.

The total strength of 48 conservation guards from the society have been giving full-time service for the protection of flora and fauna of the National Park. The guards too have records of recovering hand-made guns (*Gazimara)* from poachers. All kinds of anti-forest activities are directly prohibited by MMES conservation guards. Villagers sometime try to violate the Wildlife Protection Act due to their ignorance. In such situations MMES intervenes and such villagers are provided with knowledge regarding the Forest Act and the importance of conservation of nature. The aim of MMES is to win the minds of all people and make them responsible towards conservation of biodiversity.

Apart from this, MMES undertake night patrolling and camping inside the park as many areas of the park are not covered by protection camps to prevent illegal or smuggling activities. MMES currently provides services to NGOs that require assistance in the rehabilitation and conservation of Manas.

Achievements

The Manas Maozigendri Ecotourism Society has emerged as a conservation messiah in Manas National Park at a time when the park was severely depleted. The emergence of MMES marks a shift in the area's thinking from TERRORISM to TOURISM. The Society has been working tirelessly in Manas and the surrounding areas to promote sustainable conservation and responsible ecotourism.

MMES is found to be a successful model that can be adopted in other places for biodiversity conservation with community participation. MMES was successful in motivating former poachers and extremists about the need for biodiversity protection and including them in the upkeep of Manas National Park. The various achievements of MMES are:

- Tourism activities in Manas under MMES since 2005. Eco-tourism activities promoted by MMES have received wide appreciation from tourists both domestic and foreign. Engagement of local people from within the Bodo community to promote ecotourism in the region has led to the development of the society at the same time. Presently local people are taking greater interest in building lodges and home stays instead of looking for government jobs. Interested youth are enjoying their earnings from becoming tourist guides, driver, shopkeepers, weavers, waiter, manager, entrepreneurs, and so on. As local communities are being benefitted, they are extending their support to sustain the MMES initiative.
- The ecosystem of the area has been balanced and biodiversity is well managed with the initiative of MMES. Thereby, local students are getting the opportunity to do their research in Manas National Park with the help of the members of MMES. Manas is now a paradise for research enthusiasts and wildlife lovers.
- MMES has received appreciation from UNESCO World Heritage Commission – international recognition for their commendable work. For outstanding services towards biodiversity restoration in Manas, MMES has received the Anirudh Bhargava Award of Environment from INTACH in 2005; the Prestigious Amrita Devi Bishnoi Wildlife Protection Award 2006 from the Ministry of Environment & Forests, Govt. of India; Award of Excellence in Conservation 2006 from Kolkata Conservators of Nature; Award of Excellence 2012 for Outstanding Contribution to the Tourism Sector from Assam Tourism; Certificate of Appreciation from Govt. of

BTC; and Certificate of Appreciation from Tourism and Rural Development, Government of Assam, BTC on World Tourism Day 2020.

- Mr. Rustam Basumutary, who is one of the prime founder members of MMES, has received the Assam Gaurav Award on 4 January 2023

Conclusion

Community-based ecotourism initiatives like MMES are a successful model of eco-tourism practice. It has achieved sustainability in the dimensions of ecology, economy, and development of society. It has made significant contributions to the Bodo community's improvement in standard of living. In terms of economics, MMES' community-based ecotourism effort in Manas National Park has resulted in long-term advantages that are dispersed equitably among the host communities, which has also helped to improve the community's infrastructure on a continuous basis.

Furthermore, it has aided local people's psychological empowerment by enhancing their self-esteem and nurturing pride in their cultural and natural heritage. This occurs when ecotourism highlights the worth of the host community's natural beauty or cultural uniqueness to the public. MMES' community-based ecotourism effort has established social ties throughout the community by encouraging people to work together. It has aided political empowerment by providing a platform for citizens to express their opinions on local development concerns. Thus, it can be concluded that community-based tourism activities in Manas National Park have been benefiting the local community and biodiversity conservation and management at large.

The Maozigendri model of conservation is sustainable and beneficial for wider acceptance because every frontline staff of MMES is a volunteer of nature conservation and social upliftment. Many student leaders have come forward to volunteer for environment conservation with MMES today. Such models, if adopted, would help to improve the quality of life among the villagers while encouraging students and nature lovers to get involved actively for the noble cause of nature and society. The main motto of MMES model initiative is to encourage community participation.

Bibliography

Asmamaw, D., & Verma, A. (2013) Ecotourism for environmental conservation and community livelihoods, the case of the Bale Mountain National Park. *Ethiopia Journal of Environmental Science and Water Resources*, 2(8), 250–259.

Mochahary, H. (2019). Traditional culture of bodos and its changes. *Journal of Research in Humanities and Social Science*, 7(7), 9–13.

Goodwin, H., & Rosa, S. (2009) *Community-based tourism: A success?* University of Greenwich.

Hussain, I. (2021) Ecotourism and local resource based sustainable livelihood: A case study of Kaziranga National Park. *International Journal of Multidisciplinary Educational Research*, *10*(8), 09–13.

Kibria, A., Behie, A., Costanza, R., Groves, C., & Farrell, T. (2020). Potentials of community-based-ecotourism to improve human wellbeing in Cambodia: An application of millennium ecosystem assessment framework. *International Journal of Sustainable Development & World Ecology*, *28*, 461–472.

Kiss, A. (2004) Is community-based ecotourism a good use of biodiversity conservation funds? *Trends in Ecology and Evolution*, *19*(5), 232–237.

Ministry of Environment and Forest, Government of Assam. (2019). https://forest.assam.gov.in retrieved on date 09.01.2023.

Stronza, A., Durham, H., & William (2008) Ecotourism and conservation in the Americas. Available at www.cabi.org (accessed on 21 October 2022).

United Nations. (2001) UNEP manual for the international year of ecotourism (pp 1–18). Retrieved from www.uneptie.org/tourism/home/html: United Nations Environment Program.

Wearing, S., & Neil, J. (1999). *Ecotourism: Impacts, potentials and possibilities* (p. 163). Reed Educational and Professional Publishing Ltd.

Wood, M. E., (2002) *Ecotourism: Principles, practices and policies for sustainability*. United Nations Environment Program, Division of Technology, Industry and Economics and the International Ecotourism.

Bhattacharjee, S., & Boro, F. (2017). Significance of bailwng in the traditional belief and customs of the bodos in Assam. *Journal of the Anthropological Survey of India*, 66(1–2), 277–295. https://doi.org/10.1177/2277436X20170118

Bhattacharya, A. K., Banerjee, S., & Saksena, V. (2003). Community based ecotourism: A case study of Kerwa Van Vihar National Park Catchments, Bhopal (Madhya Pradesh). *Tourism Recreation Research*, *28*(1), 97–102.

Borah, A., & Ryngnga, P. K. (2019). Ex-situ conservation of biodiversity through responsible tourism: A study of eastern himalayan botanic ark. *International Journal of Research and Analytical Reviews (IJRAR)*, *6*, 985–992.

Britannica, T. (Editors of Encyclopaedia) (2017, July 28). *Bodo language. Encyclopedia Britannica*. https://www.britannica.com/topic/Bodo-language

Ties announces ecotourism principles revision – the International Ecotourism Society. (2015). https://ecotourism.org/news/ties-announces-ecotourism-principles-revision/

What Is Ecotourism – the International Ecotourism Society. (2015). https://ecotourism.org/what-is-ecotourism/

WWF International. (2001). *Guidelines for community-based ecotourism development*. WWF. https://wwf.panda.org/wwf_news/?12002%2FGuidelines-for-Community-based-Ecotourism-Development

16 Re-envisioning Disaster Management in Assam

Happy Baglari and Mousum Duari

Our earth is the only planet that is known to support life. Composed of 71% of water and the remaining 29% constituting the continents and islands, it is also known as the blue planet. As estimated by scientists, it is home to 8.7 million species existing in a symbiotic relationship.

Climate change refers to long-term shifts in temperatures and weather patterns. These shifts may be natural, such as through variations in the solar cycle. But since the 1800s, it was seen that human activities have been the main driver of climate change, primarily due to burning fossil fuels like coal, oil, and gas. Burning of fossil fuels generates greenhouse gas emissions that act like a blanket wrapped around the Earth, trapping the sun's heat and raising temperatures. Greenhouse gas emissions are causing climate change. Uncontrolled usage of gasoline for driving a car or coal for heating a building, clearing land and forests, and landfills for garbage are major sources of methane emissions not excluding the energy industry, transport, buildings, agriculture, and land use that are among the main emitters. The Intergovernmental Panel on Climate Change (IPCC) of the UN has reported that the Earth is now about 1.1°C warmer than it was in the late 1800s. The last decade (2011–2020) was the warmest on record.

Global frameworks and agreements, such as the Sustainable Development Goals (2015) and Paris Agreement (2015) under the United Nations Framework Convention on Climate Change (UNFCCC) aim to deliver economic benefits while improving lives and protecting the environment. Three broad categories of action are: cutting emissions, adapting to climate impacts, and financing required adjustments. Also, a switch in energy systems from fossil fuels to renewable sources like solar or wind will reduce the emissions driving climate change. While a growing coalition of countries is committing to net zero emissions by 2050, about half of emission cuts must be in place by 2030 to keep warming below 1.5°C. Fossil fuel production must decline by roughly 6% per year between 2020 and 2030 (IPCC, UN, 2022). In 2022, COP27, that is, the 27th Conference of the Parties, was held from 6 November till 20 November 2022 in Sharm El Sheikh, Egypt. It was the 27th United Nations Climate Change conference which took place under the presidency of Egyptian Minister of Foreign Affairs. The United Nations Climate Change

DOI: 10.4324/9781003504870-19

Conference COP27 discussion ended with a breakthrough agreement to provide "loss and damage" funding for vulnerable countries that were hit hard by climate-related disasters. The conference emphasized the need to strengthen systematic observations and to achieve universal coverage of early warning systems COP 28 to the United Nations Framework Convention on Climate Change (UNFCCC) took place from November 13, 2023, in Dubai, United Arab Emirates where some 85,000 participants, including more than 150 heads of state and government participated. It marked the conclusion of the first "global stocktake" of the world's efforts to address climate change under the Paris Agreement. The conference focussed on accelerating the transition away from fossil fuels and reducing emissions before 2030 and integrating nature, people, livelihoods, and well-being into climate action strategies. COP 28 concluded with a renewed sense of urgency for global climate action and highlighted the need for concrete steps and increased financial support for developing countries.

Variation in climate, with its high and low temperatures, can be linked to significant psychological impacts in individuals and on communities. One significant impact due to change in climate, especially during colder, darker seasons, contributes to the development of seasonal affective disorder (SAD) which occurs due to the colder temperatures and reduced exposure to sunlight, and it disrupts the body's internal clock which can cause feelings of depression, lethargy, and irritability. Another issue caused by prolonged exposure to high temperature is psychological distress and anxiety, along with displacement of individual's property and lives due to climate-related events such as hurricanes, floods, wildfires, and droughts, which can lead to trauma. Fluctuation in temperature can disrupt sleep patterns, which can in turn contribute to mood related issues. Overall, there is an immense impact of climate and temperature on physical and psychological health.

India

India's first-ever climate change assessment report was carried out in 2020 by Krishnan et al. and reported in the book *Assessment of Climate Change over the Indian Region*, which has revealed that the country's average temperature is expected to rise by 4.4 degrees Celsius by the end of the year 2100. In India, climate change is already affecting human health, wildlife, food production, clean water access, and the economy at large (Krishnan et al., 2020).

It is noteworthy that India is one of the few countries where environmental education is compulsory in formal education. But 65% of the Indian population remains unaware of climate change as a problem, according to Yale University. Of the remaining 35% – mostly educated people – 80% see it as a serious threat (Thomson Returns Foundation News, 13th July, 2021). A recent survey by an educational company called Brainly found that 79% of Indian students feel it's important to study climate change and environmental conservation (Brainly, 2023). India's first endeavour to include environmental education in formal education was made under the National Policy on Education in 1986. The National Curriculum Framework, in 2005, further stressed the integration of environmental issues and recommended project-based learning. In 2016, the University Grants Commission introduced a six-month course on environment studies, making it mandatory for undergraduates from all disciplines, but the themes introduced in the curriculum are far away from the immediate concerns of climate change.

The other important measures taken up by the Indian government towards sustainable development are, firstly, through using the 3-R approach of reduce, recycle, and reuse. Secondly, promoting and developing sustainable lifestyle. Thirdly, putting focus on the appropriate use of technology, that which is eco-friendly and culturally sustainable.

North Eastern States

The North East of India is home to about 45 million people and accounts for almost 8% of the nation's territory (Dikshit and Dikshit, 2014). The North Eastern region has been in focus for its high diversity, and this region has been considered a prime concern by leading conservation agencies of the world. While World Wildlife Fund has identified the entire Eastern Himalayas as a priority, Global 200 Eco region, Conservation International has upscaled the Eastern Himalayan Hotspot which initially covered the states of Arunachal Pradesh, Sikkim, Darjeeling Hills, Bhutan, and South China to the Indo Burma Hotspot (Myers et al., 2000).

According to Toky and Ramakrishnan (1983), approximately 50% of India's forest cover is in the North East Region. However, increasing population pressure and the extractive process of development, particularly in the last two decades, has begun to exact a heavy toll on the environment (I. Baruah, 2002).

Assam

Assam, a state in North Eastern India, is known for Assam tea and Assam silk. Assam also shares international borders with Bhutan. In the north of it there are the Eastern Himalayas along the Brahmaputra valley and to the south is the Barak River valley. Approximately, Assam covers an area of 78,438 km^2 (30,285 sq mi).

Assam, being gateway to the North East states, is a mixture of diverse cultures and the people here are an intermixture of numerous racial stocks such as Mongoloid, Indo-Burmese, Indo-Iranian, and Aryan. The state also consists of a large number of tribes such as Bodo, Sonowal Kachari, Karbi, Miri, Mishimi, Rabha, and others, each unique in its own traditions, culture, and dress. Most tribes have their own language, although the official language is Assamese. There are two additional official languages, Bengali and Bodo.

Further, 'Muga', which is exalted in Assam and known to the world as 'Assam Silk', is a unique regional entity found only in North East India. The state is also a major producer of Eri silk, another icon of the region. In addition to its traditional uses as fabric, silk has applications in skin care, tissue engineering, and more. At Biodiverse 2018, a conference organized by IIT Guwahati, in his address, Kartik Neog (a scientist with the Central Muga Eri Research & Training Institute (CMER&TI)), said that:

> The Muga silkworm performs best in the temperature range of 25–27°C and humidity of 75–85 percent. Generally, a temperature range minimum of 15°C and a maximum of 33°C is considered as the range of temperature suitable for Muga silkworm rearing. Beyond these two ranges, mortality of Muga silkworm increases.

Neog further added that despite being awarded a geographical indication designation marking the uniqueness of local silk, silkworms and related industries remain vulnerable. The widespread use of pesticides, disease and pests, and global warming in nearby tea fields threaten the survival of silkworms.

The diversity of Assam does not just end here. Assam was the first site in Asia for drilling for oil. It is home to the one-horned Indian rhinoceros, along with various species of Asiatic birds. The economy of Assam is aided by wildlife tourism to Kaziranga National Park and Manas National Park, which are recognized as World Heritage Sites.

Addressing the economic and psychological impacts of climate change requires comprehensive strategies that encompass mitigation efforts to reduce greenhouse gas emissions, adaptation measures to build resilience to climate-related risks, and support systems to address the psychological well-being of individuals and communities affected by climate change.

Vulnerability to Disasters

Disasters are sudden and cause great damage, loss, or destruction of properties and life. Disasters are categorized based on their cause, that is, natural and man-made disasters. Natural disasters are caused by nature, and men have no control over them such as earthquakes, tsunamis, floods, landslides, hurricanes, wildfires, droughts, and volcanic eruptions. Such disasters cause massive loss of life, property, and many other miseries.

Man-made disasters are caused by human beings. Some examples of man-made disasters include hazardous material spills, explosions, chemical or biological attacks, nuclear blasts, traffic collisions such as train accidents, plane crashes, and so on. Most of these disasters are in the form of accidents (except attacks) and cause deaths, injuries, and loss of property.

Assam is a multi-hazard state prone to floods, earthquake, storms, and landslide besides man-made disasters. The state also faces acute floods and soil erosion. Assam has a history of disasters ranging from large earthquakes to severe floods. The state has experienced two major earthquakes in 1897 and 1950.

Thunderstorms known as '*Bordoicila*' are frequent in Assam during the months from April to June. The wild wind is accompanied by heavy downpour that starts from the end of March to the beginning of April.

Man-made disasters usually result from human activities such as coal mining, oil, and gas drilling. Coal mining exerts a long-lasting impact on landscape, ecosystem, and socio-cultural-economic considerations. Margherita, known for coal mining, is in the North Eastern coal fields of Assam. The mining has caused significant forest and topsoil loss, acid mine drainage, and soil and water pollution in that area.

With all the disastrous effects elaborated here, Ecosystem restoration can be a tool for protecting biological diversity on the planet, and that restoration of an ecosystem requires the re-creation of prior structure and function (Cairns, 1980), including ecosystem services (Cairns, 1986).

Evidence-Based Havoc over Natural Habitat

Floods in Assam

Floods are one of the most common natural disasters. Flood can be defined as the rising and overflowing of a body of water onto normally dry land. Flood occurs in Assam due to heavy monsoon rains. The flood-prone area of the state as assessed by the Rastriya Barh Ayog (RBA) is 31.05 lakh hectares against the total area of the state of 78.523 lakh hectares, that is, about 39.58% of the total land area of Assam This is about 9.40% of the country's total flood-prone area. According to Government of Assam Water Resources, average annual area affected by flood is 9.31 lakh hectares. Assam faced major floods in 1954, 1962, 1972, 1977, 1984, 1988, 1998, 2002, 2004, and 2012. Almost every year three to four waves of flood ravage the flood-prone areas of Assam. Average annual loss due to flood in Assam is to the tune of Rs. 200.00 Crores.

The recurrent flood is generally followed by severe impact on psychosocial health and socio-economic vulnerability. The flood combined with river erosion has significant impacts each year which destroy human life, crops, and other livelihood. Assam is mainly dependent on agriculture not only for its economy but for overall development. If proper actions are not taken against flood, then it is likely to affect the future generations of the state (Gogoi, 2016).

According to National Disaster Response Force, Ministry of Home Affairs, Govt. of India, the state has been facing severe floods and landslides since April 6 2022, due to pre-monsoon showers. As per Central Water Commission,rivers, namely Kopili, Brahmaputra, Burhidihing, and Disang have flowed above danger level, Barak (at BP Ghat), Karimganj (at AP Ghat), and in Cachar and Kushiyara (at Karimganj) were flowing above danger level.

Another report, "Flooding in India's North East Reveals Weakness of the Act East Policy" by A. Choudhury (2022), stated that North East India, especially Assam and Meghalaya, encountered one of its most terrible floods in the year 2022, which left an unfading imprint on lives.

Oil field Blowout in Assam

There are several man-made disasters other than the oil field blowouts. One such is Bhopal gas tragedy which is considered as the world's worst industrial disaster. In Assam, between the twin crises of COVID-19 and ongoing flood, there was the massive disaster at the Baghjan oil field.

Oil fields in North East India

The major oil fields in North East India are those of the Brahmaputra valley in Assam and its neighbouring areas including Arunachal Pradesh, Nagaland, Meghalaya, Tripura, Manipur, and Mizoram. Assam is the oldest oil-producing state in India. The main oil-bearing strata extend for 320 km in upper Assam along the Brahmaputra valley. Some of the important oil fields of Assam are

- The Digboi field
- The Naharkatiya field
- The Moran-Hugrijan field
- The Baghjan field

The Baghjan Disaster

The Baghjan oil field has 21 active wells, out of which four produce natural gas while the remaining produce oil. On 27 May 2020, the Baghjan oil field incident in Assam, India represented a significant tragedy with wide-ranging impacts when a massive and uncontrollable natural gas fire broke out in the Baghjan oil field, operated by Oil India Limited. The blowout occurred at Well No. 5, which was established in 2006 and produced natural gas. It was drilled down to a depth of 3,870 meters and was producing between 80,000 and 100,000 standard cubic meters of gas per day before the leak. In the morning hours of 27 May 2020, residents of Baghjan reported hearing a loud sound from the oil field and the sound of gushing of natural gas from the well continued to produce an audible sound for a number of days. Although production was stopped at Well No. 5, during performing service to the well, workers noticed that gas was escaping from the capped well, and the blowout occurred because of that. Following the blowout, 1610 local families, consisting of approximately 3000 persons were evacuated to relief camps. A safety belt of 1.5 km radius around the well was established. At last, the blowout was "killed" on 15 November 2020, after 173 days from the blowout, with a technique called snubbing. Subsequently, Well No. 5 was abandoned on 3 December 2020. The unfortunate outburst took the lives of many firefighters and enormously affected the biodiversity and the lives of those residing nearby. It has been estimated that the outburst has not only affected the site but also the neighbouring areas, the people, and the ecosystem.

Gogoi et al. (2022) conducted a case study on the Baghjan tragedy and stated that on 18 January 2020, the Central Government discharged all oil and gas firms operating exploratory drilling from legal requirements for environmental clearances. Before the leak, in 2020, the region witnessed a protest over the announcement from the government of India that they had authorized Oil India Limited to explore the area under Dibru-Saikhowa National Park for hydrocarbons. An investigation by Assam's State Pollution Control Board found that Oil India had begun drilling at the Baghjan Oil field without procuring prior environmental clearance following legal requirements under the Water (Prevention of Control of Pollution) Act, 1974, Air (Prevention of Control of Pollution) Act, 1981 and Environmental Protection Act, 1986. Oil India Limited had also failed to conduct mandatory public hearings before beginning drilling. In 2013, the Baghjan oil field was inspected by a team from the National Board for Wildlife, which criticized the government of India for approving formally the violation of wildlife norms after Oil India Limited had already begun operating in an eco-sensitive zone.

The State Pollution Control Board officials have stated that the local crops and plants had been affected by condensation of the leaked gas. Wind conditions had carried the leaked gas towards the Dibru-Saikhowa National Park as well. Wildlife Division officials noted that immediately after the leak there was heavy rainfall due to which the local waterbodies had been contaminated by condensate from the leak. Following the leak, the State Forest Department had asked Oil India Limited to account for its actions after there were reports of dead fish in local water bodies. On 5 June 2020, while commemorating World Environment Day, residents protested near the Maguri-Motapung Beel wetland, calling for the protection of the ecologically sensitive Dibru-Saikhowa National Park, and called for compensation for their loss of livelihoods because of the leak.

Psychological Impact of Disasters

Both natural and man-made disasters cause psychological impact and pose threat to one's life. Survivors face difficulty in handling the effects of disasters. The difficulty will be in the form of survivors' unfamiliarity with the basic principles of psychosocial support, shortage of relevant experts, inadequate training, losing close ones, nightmares, and flashback of the trauma that one has gone through.

Every year most parts of the North Eastern region encounter heavy rainfall and Assam, with Barak and Brahmaputra valleys being the prominent flood plains, is affected the most. This brings havoc not only to the lives of people but also animals, properties, plants, roads, infrastructure, and a lot more. Moreover, there are huge psychosocial effects of such disasters on one's life. Many lose their loved ones and their properties, which can be traumatizing and can have lasting psychological effects as well.

Gogoi et al. (2022) in their report stated that the effects and aftereffects of the Baghjan incident is still deeply engraved on the area and its inhabitants. The residents who were moved to emergency camps now complain of fear and hearing damage and respiratory diseases. The authors, moreover, conducted unstructured interviews utilizing an observatory-conversational method. With the consent of the participants, the conversations were audio-recorded with open-ended questions. The data and information generated from these methods addressed the after issues or the psychological impacts on the inhabitants. The participants stated of having nightmares and insomnia worrying about their family and kids' future.

Analysis and Discussion

To delve deep into the gas-leak disaster, secondary data was collected through systematic study of various peer review, Scopus, and Web of Science research papers. Extensive literature has been surveyed and analyzed, and various articles, case studies, and newspaper articles have been reviewed.

Some plausible reasons for the occurrence of the disaster have been associated with certain human factors like lack of attention, poor workmanship, and aging workforce as well as bad maintenance of the machines and morpho-tectonic factors which may have led to the imbalance of pressure in the oil well causing the sudden blowouts. According to Indian Express News report, one of the temporary steps taken by the authorities was building a water reservoir through pipelines from the Dangori river near the well to control the blowout. Challenges were faced as the gas leak at Baghjan was still not under control, and it was continuously leaking, as the limited space and non-availability of open space above the well-head posed a huge challenge in the placement of the blow out preventer.

An investigation by a committee appointed by the National Green Tribunal revealed in November 2020 that Oil India Limited had failed to obtain legally required clearances to operate the oil field at Baghjan and had failed to comply with provisions of environmental laws as well as with internal safety procedures concerning drilling at the oil field. The findings reiterated by the Assam State Pollution Control Board, which had also reported that Oil India Limited did not obtain the necessary clearances before drilling.

In terms of issues and concerns, various lives have been destroyed and families have been separated for days during the Baghjan tragedy. Most people had to stay in relief camps. It was a sudden shock for all of them. All their belongings back at home were burnt to ashes. Fields were completely destroyed; for most it was the sole source of their income. Though with the compensation people are rebuilding the houses and people have cultivated rice, its productivity is doubtful. All of this had made them more worried about their future. These depict that the tragedy has not only affected the lives and properties of people but their resilience capacity, apprehending if they could ever come back to a normal state and build back everything as earlier. The gas leak has caused the deaths of Gangetic dolphins and varieties of fish. The number of birds has decreased because they have flown away to distant places.

From the preceding observations it can be understood that disasters have their greatest impact at the local level which is now becoming more complex and diversified due to the intersectionality between natural and human-induced disasters; vulnerability; risk perception; resource distribution; social categorizations such as class, caste, and gender; and government fronts. Therefore, the involvement and mobilization of the affected local communities has always, and now more than ever, been an essential component of the whole cycle of disaster management.

Communities are a source of strength, experience, and knowledge of living and adapting to the prevailing disasters in their own different ways. There are references to many incidences when local people predicted impending earthquakes or cyclones by observing natural signs such as animal behaviour, changes in characteristics of ground water sources, and abnormal weather phenomena. But in the case of Baghjan the local communities were not prepared for such a disaster as they were not accustomed to or warned of such unforeseen incidents. They were never given any training or awareness on how they should practice resilience and preparedness, being communities that reside among active oil wells and natural gas outlets.

Another concern is the annual flood. There has always been debate whether flood is a natural disaster or man-made. No doubt the entire North East is a flood-prone zone due to their geographical and topographical characteristics, making it one of the highest rainfall zones in the world. Brahmaputra is considered as an aggressive river which flows at a rapid speed from the Himalaya and gets connected to other tributaries, such that when all these waters get united it results in flooding. As mentioned, it cannot be fully considered as a natural disaster. Since the 1950s, people have been coming up with temporary solutions to flooding such as building embankments.

Re-envisioning Disaster Management

The Baghjan gas leak and the annual floods not only affected the lives and properties of people but challenged the psychosocial abilities of the people to bounce back to a normal state. Traumatizing events like these need to be dealt in ways that are indigenous to the region. Some measures are as follows:

- Management, Protection and Reestablishment: Restoring natural habitats involves more than just planting trees or reintroducing species. It requires a comprehensive approach that considers both abiotic (non-living) and biotic (living) factors to return ecosystems to their historical functioning levels. Here is an elaboration on how management, protection, and reestablishment can achieve this:

Understanding Historical Conditions:

a Before initiating any restoration efforts, it is crucial to understand the historical conditions of the ecosystem, including its natural vegetation, soil composition, hydrology, and species interactions.
b Historical data, including records, photographs, and ecological surveys, can provide insights into the ecosystem's original state and guide restoration goals.

Assessing Human Impacts:

a Human activities such as deforestation, urbanization, pollution, and invasive species introduction often disrupt natural habitats, altering abiotic and biotic factors.
b Assessing the extent and nature of human impacts helps identify specific restoration needs and prioritize management and protection strategies.

Management of Abiotic Factors:

a Abiotic factors such as soil quality, water availability, and nutrient levels play a critical role in habitat restoration.
b Soil erosion control measures, erosion control blankets, and revegetation with native plant species help stabilize soil and restore soil structure.
c Hydrological restoration techniques such as restoring natural water flow patterns, constructing wetlands, and implementing erosion control measures can improve water quality and quantity.
d Soil amendments, such as adding organic matter or minerals, may be necessary to enhance soil fertility and support plant growth.

Protection of Biotic Factors:

a Protecting native flora and fauna from further degradation and preserving existing biodiversity are essential aspects of habitat restoration.
b Removing invasive species and controlling exotic pests help reduce competition for resources and restore the balance of native species.
c Creating wildlife corridors and protected areas facilitates the movement of species and supports population recovery.
d Implementing habitat management practices, such as controlled burns or selective logging, mimics natural disturbances and promotes habitat heterogeneity, benefiting a wide range of species.

Re-establishment of Native Flora and Fauna:

a Reintroducing native plant species that are adapted to local environmental conditions helps rebuild ecosystem structure and function.
b Establishing plant communities that mimic natural succession patterns and restoring keystone species can enhance ecosystem resilience and stability.
c Monitoring and adaptive management are crucial to evaluate restoration progress, adjust management practices, and address emerging challenges such as climate change impacts.

- Construction of Reservoirs: Constructing reservoirs on tributaries and distributaries can serve as an effective measure for flood storage and integrated water management.

Flood Storage:

a Reservoirs act as large storage basins that can capture excess water during periods of heavy rainfall or snowmelt, thus reducing downstream flood risk.
b By strategically locating reservoirs on tributaries and distributaries, where water naturally converges before flowing into main rivers, floodwaters can be temporarily stored and released in a controlled manner.
c This helps regulate the flow of water downstream, mitigating the risk of flash floods and reducing damage to downstream communities, infrastructure, and agriculture.

Water Supply Management:

a Reservoirs serve as important water storage facilities for domestic, agricultural, industrial, and municipal use.
b By capturing and storing water during periods of high flow, reservoirs ensure a reliable water supply during dry seasons or droughts when natural water sources may become scarce.
c Integrated reservoir systems on tributaries and distributaries allow for the efficient distribution of water to various users and sectors, helping meet the diverse water needs of communities and industries.

Hydropower Generation:

a Reservoirs can be equipped with hydropower facilities to generate electricity from the stored water's potential energy.
b By harnessing the energy of flowing water, hydropower plants contribute to renewable energy production, reducing reliance on fossil fuels and mitigating greenhouse gas emissions.
c Integrating hydropower generation with flood storage reservoirs allows for a dual-purpose infrastructure that provides both flood control and clean energy generation.
d Reservoirs often create recreational opportunities such as boating, fishing, swimming, and wildlife viewing, which can stimulate local economies through tourism and outdoor recreation industries.
e Well-designed reservoirs with appropriate shoreline facilities and amenities can attract visitors and enhance the quality of life for nearby communities.

- Resilience Building: The potential to bounce back from any stage of adversity is resilience, especially in the younger generations as they are the future hope and the creators. Being disaster resilient offers much promise as a torchbearer for the promotion of disaster risk management.

Christine A. Pade and Kathleen A. Mooney have discovered a four-step cognitive behavioural therapy to help clients build positive qualities and strengthen personal resilience.

Their model includes four steps to resilience:

Step 1: Search for strength: By search for strength, they mean determining individual's strengths such as strategies, beliefs, and personal assets and with the use of these, one can effortlessly stimulate the positive qualities one is attempting to build, that is, resilience in this case.

Step 2: Construct a personal model of resilience (PMR): The therapist and the client (who have gone through any adversity) together designed a PMR based on the strengths associated and written down during the search phase. The PMR designed must be written in client's own words and the client generates imagery and metaphors to help make the PMR remarkable.

Step 3: Apply the PMR: After the PMR is built, the client is suggested to reflect on how this constructed PMR could assist to maintain resilience in areas of difficulty. Common challenges in problem areas are examined and written down. The therapist asks clients to look through their PMR, aiming for ideas that would help them endure at times of adversity or at times of facing obstacles and/or to accept facets of the situation that cannot be changed. The focal point of this discussion is on staying resilient in the face of difficulties rather than success in finding a solution or overcoming them.

Step 4: Practice Resilience: During the final stage of strengths-based cognitive behaviour therapy for resilience, therapist and clients devise behavioural experiments to practice resilience such as the resiliency lens which focusses on how long the client stayed resilient, what strategies helped the most, how he felt, and so forth.

- Therapeutic Garden: Introducing therapeutic gardens in educational institutions in North East India can indeed provide numerous benefits, fostering holistic learning experiences and promoting environmental consciousness among students. A therapeutic garden can serve as a valuable educational tool and contribute to positive outcomes:

Hands-on Learning Opportunities:

a Therapeutic gardens offer hands-on learning experiences where students can actively engage in planting, nurturing, and caring for plants.

b Incorporating lessons on numbers, alphabets, and language skills into gardening activities makes learning more interactive and enjoyable for students.

c Through practical activities like seed planting, watering, and observing plant growth, students develop a deeper understanding of concepts related to biology, ecology, and environmental science.

Environmental Education and Awareness:

a Therapeutic gardens provide a platform to educate students about the importance of trees, biodiversity, and environmental conservation.
b By witnessing the lifecycle of plants firsthand, students gain insights into the ecological processes, the interdependence of living organisms, and the significance of maintaining healthy ecosystems.
c Discussions on topics such as soil health, water conservation, pollination, and sustainable gardening practices raise students' awareness of environmental issues and empower them to become responsible stewards of the environment.

Cultivation of Life Skills and Values:

a Gardening fosters the development of essential life skills such as patience, responsibility, teamwork, and problem-solving.
b Students learn the value of perseverance as they witness the gradual growth and development of plants over time.
c Collaborative gardening projects encourage teamwork and cooperation among students, teaching them the importance of mutual support and collective action.

Promotion of Sustainable Practices:

a Therapeutic gardens serve as living examples of sustainable practices, demonstrating the principles of organic gardening, composting, and resource conservation.
b Through composting organic waste, conserving water, and using eco-friendly gardening techniques, students learn how to minimize their environmental footprint and contribute to a more sustainable future.
c By cultivating a culture of environmental stewardship and sustainability, therapeutic gardens inspire students to make informed choices and take actions that benefit both themselves and the planet.

Integrating therapeutic gardens into educational institutions in North East India offers a holistic approach to learning, promoting academic achievement, environmental awareness, and personal growth. By nurturing a connection to nature and instilling values of environmental stewardship, therapeutic gardens empower students to become lifelong learners and agents of positive change in their communities.

Indeed, the concepts outlined in the National Education Policy (NEP) 2020, particularly the Foundation Stage and the Preparatory Stage, align well with the principles of Therapeutic Gardens and can complement each other effectively.

- Foundation Stage (5 years):

  i The Foundation Stage, with its focus on flexible, multilevel, play-based, activity-based, and discovery-based learning, provides an ideal environment for incorporating Therapeutic Gardens into early childhood education.

ii By integrating time-tested Indian traditions and cultures into gardening activities, children learn about their cultural heritage and develop a deeper connection to nature and their surroundings.

- Preparatory Stage (3 years)

i During the Preparatory Stage, children build upon the play-, discovery-, and activity-based learning experiences from the Foundation Stage. Introducing formal classroom learning with textbooks gradually prepares them for more structured academic pursuits.

In both stages, the Therapeutic Garden concept aligns with the holistic development goals outlined in the NEP 2020, including cognitive, emotional, and socio-cultural dimensions. By integrating gardening activities into the curriculum, educators can create dynamic learning environments that foster creativity, critical thinking, and personal growth. Additionally, Therapeutic Gardens promote environmental awareness, sustainability, and a sense of connection to the natural world, aligning with the NEP's emphasis on holistic and experiential learning approaches.

Incorporating Therapeutic Gardens into early childhood and preparatory education stages enhances the quality and effectiveness of learning experiences, nurturing well-rounded individuals who are equipped with the knowledge, skills, and values needed for success in the 21st century.

There is a growing realization that community participation is the most effective element for achieving sustainable and long-term results in disaster management and risk reduction. Zubir and Amirrol (2011) have found out that many aspects of disaster risk reduction is through the participatory involvement of the local community. They discussed the application of systematic disaster management in managing emergency and post-emergency situations by adopting the role of civil society in the context of community-based approaches which are proven to be adaptive and comprehensive enough in progressing through the hard times during and after the disaster event.

The government and the existing disaster management authorities have often failed to handle every end of the various disaster dynamics, especially without the active participation of the people. There have been numerous instances where the top-down disaster management approach has not been effective according to their full potential and has created a gap between policy and practice.

While incorporating these competency-building strategies, attempts are needed to promote the close relationship that humans had with nature and restore the man-environment intimacy.

Intimacy of Man and Natural Environment

Man lives in the sphere of nature. Nature always surrounds humans. According to the biophilia hypothesis, the human species evolved in the company of other forms, and continues to rely – physically, emotionally, intellectually

– on the quality of richness of their affiliations with natural diversity (Saunders 2003). A healthy and diverse natural environment is a necessary pre-condition for human lives of satisfaction and fulfilment (Kellert and Wilson 1993). This suggests that man can evolve through its interaction with the environment and the environment in turn largely depends on man for its sustainability. Hence, it is of utmost priority to initiate practices to restore environment which would in turn support human existence.

In fact, man has a massive influence on nature and the man-nature relationship has been characterized by overuse and imbalance. Human beings aren't just mere dwellers in the sphere of nature. Due to overuse of intelligence and thinking, man's existence has adapted nature and further made all sorts of incursions into nature. Man has consistently transformed nature's wealth into the present-day economic, social, and cultural life of modern society. For instance, man has exploited nature to generate electricity and then utilized the electrical power to serve his interest and his community.

Generally, nature always fights back and wins. There are many ways in which nature has fought back: one such is climate change. Man hasn't just transformed a variety of flora and fauna to various climatic conditions, but has further altered the climate of his habitat.

Extinction of species, exploitation of resources, and global climate change are the greatest threat to humankind, which was soon realized by Mr. Jadav Molai Payeng, from Majuli in Assam. He was honoured as "Forest Man of India" by Jawaharlal Nehru University on 22 April 2012. His life story dates to 1979, age 16, when he learned that a huge number of snakes had died due to extreme heat, and after floods these were washed onto the sandbar. He wanted to do something about it, so he decided to plant around 20 bamboo seedlings on the sandbar of the Brahmaputra. He turned 550 acres of barren land into a lush green forest which has been named as "Molai Forest" after him. It was a prolonged approach as planting trees was time consuming, but the forest grew dense as the trees started providing seeds themselves, which in turn, increased the number of inhabitants. Soon, the forest was filled with hundreds of species of birds; with deer, rhinos, and tigers; and even included a herd of elephants straying into his forest three months out of the year. His initiative has not just turned a barren land into a lush green forest but became home to many species. With this, the grassroot-level implementation through conservation strategies have been proved to be a successful way to conserve environment.

With the aim of re-establishing the environment, a theory called Kaplan's attention restoration theory (ART) has been evolved, which brings out the importance of the environment and its restoration not just for supporting life but to bring peace of mind. The theory was developed and propagated by Stephen and Rachel Kaplan in the late 1980s and early 1990s. In short, this theory suggests that "mental fatigue and concentration can be improved by time spent in or looking at nature." The capacity of the brain to focus on a specific stimulus or task is limited and results in directed attention fatigue. The ART intends to say that the brain functions more effortlessly when exposed to the natural environment, thereby permitting it to retrieve and replenish its directed attention capacity.

According to Kaplan, the natural environment must have four properties to provide this restorative effect:

- Extent (the scope to feel immersed in the environment)
- Being away (providing an escape from habitual activities)
- Soft fascination (aspects of the environment that capture attention effortlessly)
- Compatibility (individuals must want to be exposed to and appreciate the environment)

This theory is relatable with the Japanese term "Shirin-Yoku" which translates to "forest bathing" or "absorbing the forest atmosphere" which was created by the Japanese Ministry of Agriculture, Forestry, and Fisheries in 1982. The goal of forest bathing is to live in the present moment while immersing the senses in the sights and sounds of a natural setting. Qing Li, a Japanese medical doctor and researcher, has found that spending time in a forest can reduce stress, anxiety, depression, and anger; strengthen the immune system; improve cardiovascular and metabolic health; and boost overall well-being. It underscores the importance of a green environment to the overall physical and psychological state of living beings. Keeping in view the current damages and with the motive to restore nature every year, world environment day is celebrated with the themes like "Ecosystem Restoration" and "Only One Earth," and each country takes part every year to globally host the day. It is time to focus on restoring not only the natural environment but restoring the intimacy of humans with nature.

Conclusion

According to the Environment Protection Act of 1990, environment consists of all, or any of the media – the air, water, and land. The National Education Policy of 1986 states that it is the basic need to develop consciousness regarding the environment in children, youth, older people, and all the citizens in every field. Survival of an organism is possible with the existence of natural habitat, a habitat inclusive of a symbiotic relationship between all living and non-living organisms.

The natural environment is precious and there is the utmost need to protect it at any cost. It is high time to adopt a new strategy that will involve vulnerable people directly in the planning and implementation of mitigation, preparedness, response, and recovery measures. Community collaboration is gaining significance as communities are becoming increasingly vocal and keen for policymakers to hear the voice of their experience. When the communities are included in the entire process involving sequential stages of disaster management, a decentralized, bottom-up, intensive, micro, and local-level approach comes into being which decreases the gap between the affected community and the existing institutionalized disaster management frameworks. Incorporating indigenous knowledge and practices about disaster preparedness, early warning systems, resilience, and risk reduction can be considered one of the significant ways of community-based frameworks. Therefore, community collaboration needs to be prioritized to increase the preparedness of the residents. It will go a long way in restoring the man-environment intimacy as well.

References

Baruah, I. (2002). Borders, ethnicity, and traditions: A passage to the north east of India. *Traditional Dwellings and Settlements Review*, 14(1), 36–37.
Brainly (2023). 79% of Indian students feel studying climate change and environmental conservation is important: Survey. *India Today*. https://www.indiatoday.in/education-today/latest-studies/story/79-of-indian-students-feel-that-it-s-important-to-study-climate-change-and-environmental-conservation-survey-1814708-2021-06-14 (accessed on 19 July 2023).
Choudhury, A. (2022). Flooding in India's northeast reveals weakness of the act east policy. https://thediplomat.com/2022/07/flooding-in-indias-northeast-reveals-weakness-of-the-act-east-policy/ (accessed on 19 July 2024)
Climate Change 2022: Impacts, adaptation and vulnerability. IPCC. http://www.ipcc.ch/report/sixth-assessment-report-working-group-ii/
Dikshit, K. R., Dikshit, J. K. (2014). *North-East India: Land, people and economy*. Springer.
Gogoi, M. (2016). Flood disaster in Assam: Socio-economic vulnerability and control measures. *SAJMS*, 3, 147–158.
Gogoi, S., Baruah, B., & Chetia, A. (2022, December). Baghjan fire: A case study of the '2020 Assam Gas and Oil Leak' at Baghjan Oil Field, Tinsukia, Assam. In *5th World Congress on Disaster Management: Volume II* (pp. 483–497). Routledge.
Kellert, S. R., & Wilson, E. O. (1993). *The biophilia hypothesis*. Harvard University Press
Krishnan, R., Sanjay, J., Gnanaseelan, C., Mujumdar, M., Kulkarni, A., & Chakraborty, S. (2020). Assessment of climate change over the Indian region: A report of the ministry of earth sciences (MOES). *Government of India* (p. 226). Springer Nature.
Myers, N., Mittermier, R. A., Mittermier, C. G., Da Fonseca, G. A. B., & Kent, J. (2000). Biodiversity hotspots for conservation priorities. *Nature*, 40, 853–858.
Saunders, C. D. (2003). The emerging field of conservation psychology. *Human Ecology Review*, 10(2), 137–149. http://www.jstor.org/stable/24706965
Toky, O. P., & Ramakrishnan, P. S. (1983). Secondary succession following slash and burn agriculture in North Eastern India: I. Biomass, litterfall and productivity. *The Journal of Ecology*, 71, 735–745.
Zubir, S. S., & Amirrol, H. (2011). Disaster risk reduction through community participation. *WIT Transactions on Ecology and the Environment*, 148, 195–206.

17 Conclusion

The Way Forward

Indranee Phookan Borooah, Sabiha Alam Choudhury and Bidita Das

Nature has been particularly kind to India's North East, bestowing it with bountiful greenery, water, and fertile soil. These together have ensured a life of sufficiency not only for its human inhabitants but for all living things, be it flora or fauna. Therefore, conservation of nature in the states of North East India is of critical importance due to the region's rich biodiversity, unique ecosystems, and cultural significance.

As evinced from the chapters in this book, climate change threatens this biodiversity through habitat loss, altered precipitation patterns, and increased incidences of various extreme weather events, including cyclones, landslides, and flash floods. Many indigenous communities in North East India depend directly on natural resources for their livelihood. Climate change poses additional challenges to these communities, affecting their traditional practices, food security, and cultural heritage.

Continued research, monitoring, and adaptation efforts are essential to better understand and address the specific impacts of climate change on North East India and to build resilience in the region's communities and ecosystems. Research has already demonstrated that it is the emotional bonding with the environment which can ensure pro-environment behavior as reinforced by the study on the adolescents of Aizawl, Mizoram. The ethnic communities of the region are replete with cultural practices, rituals, folktales, celebrations, and more, which could be resurrected and utilized to develop an intimacy with the environment in the younger generation for a sustainable future. The deep traditional knowledge regarding sustainable utilization of nature and its resources are embedded in the practices of many tribes and these have immense implications for conservation of the environment as is revealed in the various chapters on environmental intimacies.

That such a pathway is effective has been shown by the many projects written about in the book, where the intimate relationship of the ethnic groups in different states have been appealed to for conserving the environment yet sustaining the local people. Engaging these communities in conservation efforts helps promote sustainable practices and ensures the long-term success of conservation initiatives.

DOI: 10.4324/9781003504870-20

It is more than time that the last frontier of 'green environment' of India is left out from indiscriminate exploitation and pillage as was revealed by the blowout in the Baghjan oilfield of Oil India Limited. Leave the Earth alone to heal and revive! The forests, hills, rivers, and springs belong to the community. To ensure conservation, work with the community and listen to their wisdom.

By implementing these measures and fostering collaboration among stakeholders, the seven states of North East India can work towards achieving their conservation goals, protecting their rich natural heritage, and ensuring the well-being of present and future generations.

Index

Pages in *italics* refer to figures and pages in **bold** refer to tables.

For Product Safety Concerns and Information please contact our EU representative GPSR@taylorandfrancis.com
Taylor & Francis Verlag GmbH, Kaufingerstraße 24, 80331 München, Germany

www.ingramcontent.com/pod-product-compliance
Ingram Content Group UK Ltd.
Pitfield, Milton Keynes, MK11 3LW, UK
UKHW020149200926
475702UK00016B/2625
9781032825076